GOD, THE DEVIL, AND DARWIN

GOD, THE DEVIL, AND DARWIN

A CRITIQUE OF INTELLIGENT DESIGN THEORY

Niall Shanks

2004

OXFORD
UNIVERSITY PRESS

Oxford New York
Auckland Bangkok Buenos Aires Cape Town Chennai
Dar es Salaam Delhi Hong Kong Istanbul Karachi Kolkata
Kuala Lumpur Madrid Melbourne Mexico City Mumbai
Nairobi São Paulo Shanghai Taipei Tokyo Toronto

Published by Oxford University Press, Inc.
198 Madison Avenue, New York, New York 10016

www.oup.com

Library of Congress Cataloging-in-Publication Data
Shanks, Niall, 1959–
God, the devil, and Darwin : a critique of intelligent design theory / Niall Shanks.
p. cm.
Includes bibliographical references and index.
ISBN 0-19-516199-8
1. Intelligent design (Teleology) 2. Religion and science. I. Title.

BS651.S54 2003
213—dc21 2003042916

5 7 9 8 6 4

Printed in the United States of America
on acid-free paper

For My Dogs, Gnasher and Brutus

Foreword

Who owns the argument from improbability? Statistical improbability is the old standby, the creaking warhorse of all creationists from naive Bible-jocks who don't know better, to comparatively well-educated Intelligent Design "theorists," who should. There is no other creationist argument (if you discount falsehoods like "There aren't any intermediate fossils" and ignorant absurdities like "Evolution violates the second law of thermodynamics"). However superficially different they may appear, under the surface the deep structure of creationist advocacy is always the same. Something in nature—an eye, a biochemical pathway, or a cosmic constant—is too *improbable* to have come about by *chance*. Therefore it must have been *designed*. A watch demands a watchmaker. As a gratuitous bonus, the watchmaker conveniently turns out to be the Christian God (or Yahweh, or Allah, or whichever deity pervaded our particular childhood).

That this is a lousy argument has been clear ever since Hume's time, but we had to wait for Darwin to give us a satisfying replacement. Less often realized is that the argument from improbability, properly understood, backfires fatally against its main devotees. Conscientiously pursued, the statistical improbability argument leads us to a conclusion diametrically opposite to the fond hopes of the creationists. There may be good reasons for believing in a supernatural being (admittedly, I can't think of any) but the argument from design is emphatically not one of them. The argument from improbability firmly belongs to the evolutionists. Darwinian natural selection, which, contrary to a deplorably widespread misconception, is the very antithesis of a chance process, is the only known mechanism that is ultimately capable of generating improbable complexity out of simplicity. Yet it is amazing how intuitively appealing the design inference remains to huge numbers of people. Until we think it through . . . which is where Niall Shanks comes in.

Combining historical erudition with up-to-date scientific knowledge, Professor Shanks casts a clear philosopher's eye on the murky underworld inhabited by the "intelligent design" gang and their "wedge" strategy (which is every bit as creepy as it sounds) and explains, simply and logically, why they are wrong and evolution is right. Chapter follows chapter in logical sequence, moving from history through biology to cosmology, and ending with a cogent and perceptive analysis of the underlying motivations and social manipulation techniques of modern creationists, including especially the "Intelligent Design" subspecies of creationists.

Intelligent design "theory" (ID) has none of the innocent charm of old-style, revival-tent creationism. Sophistry dresses the venerable watchmaker up in two cloaks of ersatz novelty: "irreducible complexity" and "specified complexity," both wrongly attributed to recent ID authors but both much older. "Irreducible complexity" is nothing more than the familiar "What is the use of half an eye?" argument, even if it is now applied at the biochemical or the cellular level. And "specified complexity" just takes care of the point that any old haphazard pattern is as improbable as any other, *with hindsight.* A heap of detached watch parts tossed in a box is, with hindsight, as improbable as a fully functioning, genuinely complicated watch. As I put it in *The Blind Watchmaker*, "complicated things have some quality, *specifiable in advance,* that is highly unlikely to have been acquired by random chance alone. In the case of living things, the quality that is specified in advance is, in some sense, 'proficiency'; either proficiency in a particular ability such as flying, as an aero-engineer might admire it; or proficiency in something more general, such as the ability to stave off death. . . ."

Darwinism and design are both, on the face of it, candidate explanations for specified complexity. But design is fatally wounded by infinite regress. Darwinism comes through unscathed. Designers must be statistically improbable like their creations, and they therefore cannot provide an ultimate explanation. Specified complexity is the phenomenon we seek to explain. It is obviously futile to try to explain it simply by specifying even greater complexity. Darwinism really does explain it in terms of something simpler—which in turn is explained in terms of something simpler still and so on back to primeval simplicity. Design may be the temporarily correct explanation for some particular manifestation of specified complexity such as a car or a washing machine. But it can never be the ultimate explanation. Only Darwinian natural selection (as far as anyone has ever been able to discover or even credibly suggest) is even a *candidate* as an ultimate explanation.

It could conceivably turn out, as Francis Crick and Leslie Orgel once facetiously suggested, that evolution on this planet was seeded by deliberate design, in the form of bacteria sent from some distant planet in the nose cone of a space ship. But the intelligent life form on that distant planet then demands its own explanation. Sooner or later, we are going to need something better than actual design in order to explain the illusion of design. Design itself can never be an ultimate explanation. And the more statistically improbable the specified complexity under discussion, the more unlikely does any kind of design theory become, while evolution becomes correspondingly more powerfully indispensable. So all those calculations with which creationists love to browbeat their naïve audiences—the mega-astronomical odds against an entity spontaneously coming into existence by chance—are actually exercises in eloquently shooting themselves in the foot.

Worse, ID is lazy science. It poses a problem (statistical improbability) and, having recognized that the problem is difficult, it lies down under the difficulty without even *trying* to solve it. It leaps straight from the difficulty—"I can't see any solution to the problem"—to the cop-out—"Therefore a Higher Power must have done it." This would be deplorable for its idle defeatism, even if we didn't have the additional difficulty of infinite regress. To see how lazy and defeatist it is, imagine a fictional conversation between two scientists working on a hard problem, say A. L. Hodgkin and A. F. Huxley who, in real life, won the Nobel Prize for their brilliant model of the nerve impulse.

"I say, Huxley, this is a terribly difficult problem. I can't see how the nerve impulse works, can you?"

"No, Hodgkin, I can't, and these differential equations are fiendishly hard to solve. Why don't we just say give up and say that the nerve impulse propagates by Nervous Energy?"

"Excellent idea, Huxley, let's write the Letter to *Nature* now, it'll only take one line, then we can turn to something easier."

Huxley's elder brother Julian made a similar point when, long ago, he satirized vitalism as tantamount to explaining that a railway engine was propelled by *Force Locomotif.*

With the best will in the world, I can see no difference at all between *force locomotif,* or my hypothetically lazy version of Hodgkin and Huxley, and the really lazy luminaries of ID. Yet, so successful is their "wedge strategy," they are coming close to subverting the schooling of young Americans in state after state, and they are even invited to testify before congressional committees: all this while ignominiously failing to come up with a single research paper worthy of publication in a peer-reviewed journal.

Intelligent Design "theory" is pernicious nonsense which needs to be neutralized before irreparable damage is done to American education. Niall Shanks's book is a shrewd broadside in what will, I fear, be a lengthy campaign. It will not change the minds of the wedgies themselves. Nothing will do that, especially in cases where, as Shanks astutely realizes, the perceived moral, social, and political implications of a theory are judged more important than the truth of that theory. But this book will sway readers who are genuinely undecided and honestly curious. And, perhaps more importantly, it should stiffen the resolve of demoralized biology teachers, struggling to do their duty by the children in their care but threatened and intimidated by aggressive parents and school boards. Evolution should not be slipped into the curriculum timidly, apologetically or furtively. Nor should it appear late in the cycle of a child's education. For rather odd historical reasons, evolution has become a battlefield on which the forces of enlightenment confront the dark powers of ignorance and regression. Biology teachers are front-line troops, who need all the support we can give them. They, and their pupils and honest seekers after truth in general, will benefit from reading Professor Shanks's admirable book.

Richard Dawkins

Preface

A culture war is currently being waged in the United States by religious extremists who hope to turn the clock of science back to medieval times. The current assault is targeted mainly at educational institutions and science education in particular. However, it is an important fragment of a much larger rejection of the secular, rational, democratic ideals of the Enlightenment upon which the United States was founded. The chief weapon in this war is a version of creation science known as *intelligent design theory*.

The aim of intelligent design theory is to insinuate into public consciousness a new version of science—supernatural science—in which the God of Christianity (carefully not directly mentioned for legal and political reasons) is portrayed as the intelligent designer of the universe and its contents. Its central proponents are often academics with credentials from, and positions at, reputable universities. They are most assuredly not the cranks and buffoons of the church hall debating circuit of yesteryear who led the early assaults on science and science education. But the ultimate aim is the same.

The proponents of intelligent design are openly pursuing what they call a wedge strategy. First, get intelligent design taught alongside the natural sciences. Once the wedge has found this crack and gained respectability, it can be driven ever deeper to transform the end of the educational enterprise itself into a system more open with respect to its aim of religious instruction. As the wedge is driven still deeper, it is hoped that the consequent cracks will spread to other institutions, such as our legal and political institutions. At the fat end of the wedge lurks the specter of a fundamentalist Christian theocracy. This book, however, is about the thin end of the wedge: supernatural science. Ultimately, it is about two basic questions: Is intelligent design theory a scientific theory? Is there any credible evidence to support its claims?

My own experience with creationism and creation science goes back to 1996, when I had the pleasure of engaging in a public debate with Duane Gish of the Institute for Creation Research. The debate took place at East Tennessee State University, even as the Tennessee State Legislature debated the Burks-Whitson Bill to restrict the teaching of evolution in Tennessee schools. The debate in the legislature made Tennessee an international laughingstock. My debate took place about ninety miles from Dayton, Tennessee, where the infamous Scopes trial occurred, thereby showing that even those who know history are condemned to repeat it—again and again!

Teaching evolutionary biology in one of the Bible Belt's many buckles, I have had many close classroom encounters with ideas derived from creationism and creation science (including intelligent design theory). A sadly humorous account of my pedagogical trials and tribulations can be found in my essay, "Fighting for Our Sanity in Tennessee: Life on the Front Lines" (2001a). My concerns about intelligent design theory, however, run deeper than a simple worry about educational policy. Intelligent design theory represents, from the standpoints of both methodology and content, a serious challenge to the outlook of modern science itself. This is a challenge that needs to be taken seriously and not dismissed.

Accordingly, my colleague Karl Joplin and I have been engaged in a series of academic exchanges in various journals with biochemist Michael Behe, the author of *Darwin's Black Box: The Biochemical Challenge to Evolution* (see Behe 2000, 2001a; Shanks and Joplin 1999, 2000, 2001a, 2001b). I have also had an exchange with academic lawyer Phillip Johnson in the pages of the journal *Metascience* (Johnson 2000b; Shanks 2000). Johnson and Behe are the leading lights of the modern intelligent design movement in the United States (they are both senior members of the Discovery Institute), and we will meet them both again, later in this book. Needless to say, I was delighted when Peter Ohlin of Oxford University Press contacted me in the spring of 2002 to invite me to write a book about intelligent design theory.

In writing this book, I had the help of several friends and colleagues. First and foremost, I must give a special note of thanks to Professor Richard Dawkins, who kindly read the manuscript and honored me by writing the foreword to this volume. I must also thank my good friend Otis Dudley Duncan, who was a source of inspiration and constructive criticism throughout this project. Dudley read by night what I wrote by day, and in this way I got a much better first draft than I deserved.

I also offer my thanks to the following friends and colleagues who read fragments of the manuscript or had valuable discussions with me: David Sharp, George Gale, David Close, Steve Karsai, Dan Johnson, Rebecca Pyles, Jim Stewart, Bob Gardner, Keith Green, Bev Smith, Mark Giroux, Don Luttermoser, Hugh LaFollette, Rebecca Hanrahan, Marie Graves, Matt Young, Taner Edis, John Hardwig, Massimo Pigliucci, and Mark Perakh. I have also benefited from many helpful discussions with members of the Scirel (science and religion) discussion group organized by Jeff Wardeska here at East Tennessee State University. I am also grateful to Julia Wade and the members of the adult Sunday school at First Presbyterian Church in Elizabethton, Tennessee. These good people made an unbeliever welcome and kindly commented on a series of lectures I gave on these matters in the long, hot summer of 2002.

I would also like to give a special note of thanks to my friend and long-time collaborator, Karl Joplin, with whom I have authored several essays critical of intelligent design theory. Karl and I have taught classes together here in Tennessee, where the issues raised in this book have a special life of their own. Finally, I would like to thank Peter Ohlin at Oxford University Press for all his help in bringing this project to fruition.

Contents

GOD, THE DEVIL, AND DARWIN

Introduction

The Many Designs of the Intelligent Design Movement

Of God, the Devil, and Darwin, we have really good scientific evidence for the existence of only Darwin. Religious extremists, however, see Darwin's work (and subsequent developments in evolutionary biology) as the inspired work of the Devil, and a larger number of Christians, not so extreme in their views, claim to see in nature evidence of providential intelligent design by God.

The systematic study of nature with a view to making discoveries about God was known in the eighteenth century as natural theology. In the last half of the twentieth century, this enterprise, coupled with a literalist interpretation of the Bible as a true and accurate account of natural history and its beginnings, came to be known as creation science.

Yet in the process of becoming creation science, natural theology has mutated and evolved into a grim parody of itself. Where the natural theologians of old were in awe of the grandeur of nature, reveled in the discoveries of natural science, and saw the Book of Nature as a supplementary volume to the Book of God, the contemporary creation scientist feels compelled to substitute for the Book of Nature as we now know it a grotesque work of science fiction and fantasy, so that consistency may be maintained between preferred interpretations of the two books. The dangers here were recognized long ago, for as natural theologian Thomas Burnet (1635–1715) pointed out, "Tis a dangerous thing to ingage the authority of Scripture in disputes about the Natural World, in opposition to reason lest Time, which brings all things to light, should discover that to be evidently false which *we had made Scripture to assert*" ([1691] 1965, 16, my italics).

Following Burnet's lead, it is worth pointing out right here that one way in which we make Scripture—or any other text, for that matter—assert things is through interpretation. Biblical literalists might claim that they are reading the Bible the one true way that God intended it to be read, but merely saying this does not make it so. Many of the creationists who claim to be literalists actually have little more than a crude interpretation of the King James Version of the Bible, itself an interpretation of earlier writings and one that reflects the experiences of its seventeenth-century English authors. Yet even if one moves beyond the seventeenth century to the earliest surviving biblical writings, they still require interpretation. It is the reader who renders writings meaningful. Were Adam and Eve literally created together, as told in Genesis 1, or was Adam literally created first, and then Eve later, as told in Genesis 2? In the end, it really is all a matter about what we make Scripture assert. Decisions have to be made, and this process includes the decision to attach the stamp of divine authority to interpretations of the text that one finds congenial.

Politics and Religious Fundamentalism

The contemporary attacks on secular science and secular science education are fragments of a larger rejection of the secularism that has come to pervade modern democratic societies in the West. Though the United States is rightly considered the home of creation science, creationists have gained significant footholds outside the United States in countries such as Australia, Canada, and the United Kingdom. Indeed, the last three decades of the twentieth century have witnessed a massive global resurgence in religious fundamentalism of all stripes. While we in the West readily point a finger at Islamic fundamentalism, we all too readily downplay the Christian fundamentalism in our own midst. The social and political consequences of religious fundamentalism can be enormous, as evidenced by the plight of Iranians under the ayatollahs, the Israelis and Palestinians, the

Afghans under the Taliban, Protestants and Catholics at each others' throats in Northern Ireland, and campaigns of terror and intimidation waged against women's centers here in the United States.

Closer to home, there are growing concerns that the inability of the United States to formulate a rational foreign policy with respect to the Middle East reflects, in no small measure, pressure from Christian extremists who believe that support for the Israelis will accelerate the return of Christ. Dispensationalist theology, dating back to John Nelson Darby in 1830, teaches that before Christ's return, there will be a war in the Middle East against the restored nation of Israel. The establishment of the Jewish state in 1948 was seen as a vindication of dispensationalist claims. Now, apparently, God needs Washington's help to keep the predictions on track. However, as Doug Bandow of the Cato Institute has observed in connection with the biblical basis of this kind of end times theology:

> Curiously, there's no verse explaining that to bless the Jewish people or to be kind to them means doing whatever the secular government of a largely nonreligious people wants several thousand years later. This is junk theology at its worst. Or almost worst. Sen. James Inhofe (R-Okla) said in a speech last March: "One of the reasons I believe the spiritual door was open for an attack against the United States of America is that the policy of our government has been to ask the Israelis, and demand it with pressure, not to retaliate in a significant way against terrorist strikes that have been launched against them." (www.cato.org/dailys/06-04-02.html)

As Bandow observes, none other than Jerry Falwell has declared that God has been kind to America because "America has been kind to the Jews." After the events of 9/11, some prominent Christians blamed the attacks on the spiritual decline of the US, and suggested that God had withdrawn his protection.

For Falwell, the solution is clear: "You and I know there is not going to be any real peace in the Middle East until one day the Lord Jesus Christ sits on the Throne of David in Jerusalem" (*New York Times*, October 6, 2002). According to journalist Paul Krugman, Representative Tom DeLay, House leader and one of the most powerful people in Congress, has asserted, "Only Christianity offers a way to live in response to the realities we find in this world—only Christianity." As Krugman goes on to note: "After the Columbine school shootings, Mr. DeLay suggested that the tragedy had occurred, 'because our school systems teach our children that they are nothing but glorified apes who have evolutionized [*sic*] out of some primordial mud.' Guns don't kill people, Charles Darwin kills people" (*New York Times*, December 17, 2002). Thus we see that the current assaults on science education in the United States are really the tip of a much larger religious fundamentalist iceberg, an iceberg capable of sinking rather more than school curricula.

The consequences of religious fundamentalism are far from trivial. In recent years, we have seen how important avenues of medical research—for example, research involving stem cells, cloning, and embryonic human tissue—have been subjected to political restrictions as part of a strategy to pander to religious extremists. The result of such pandering is that crucial areas of biomedical research are now *not* being conducted in the United States. The attempts over the last three decades to restrict the teaching of evolution or to require that evidentially ungrounded theological alternatives be taught alongside it are not just peculiarities of educational policy; they are manifestations of a much deeper underlying problem generated by the resurgence of fundamentalist ideology.

Intelligent Design Theory

In the last decade of the twentieth century, creation science has spawned something called *intelligent design theory*, which preserves the core of creation science—the claim that the world and its contents result from supernatural intelligent design—while shearing away much of the biblical literalism and explicit references to God that were characteristic of the creation science from which it descends. The result has been termed *stealth creationism*—the less God is mentioned explicitly, the more likely it is that intelligent design theory will eventually fly under secular legal radar and bomb an increasingly fragile system of public education. Intelligent design theory has serious academic proponents at reputable universities, and because of clever marketing, it is having a growing influence in debates about education at local, state, and national levels. It is, in fact, a *wedge* seeking cracks in our secular democratic institutions. And intelligent design theorists themselves have made much of the metaphor of the wedge.

In this book, I explain what intelligent design theory is, where it came from, and how it is currently being presented to the public as part of a broad strategy not just to reintroduce religion into school curricula but also as a challenge to the very foundations of the modern secular state. I argue that although intelligent design theory has broad appeal to those in the sway of both Christian and Islamic fundamentalism (and as we shall see, there are some interesting ties between these two species of religious extremists), it represents a serious threat to the educational, scientific, and philosophical values of the Enlightenment that have helped to shape modern science and our modern democratic institutions. Some proponents of intelligent design theory have been quite open about this last point.

The threat to the values of the Enlightenment inherent in the intelligent design movement is particularly clear in Phillip Johnson's *Reason in the Balance: The Case against Naturalism in Science, Law and Education*. Others, more clearly identifiable than Johnson as religious extremists, have also been open about their rejection of Enlightenment values. Kent Hovind, for example, who runs Creation Science Ministries in Florida and promulgates theories favored by the antigovernment groups, maintains, "Democracy is evil and contrary to God's law" (*Intelligence Report*, Southern Poverty Law Center, Summer 2001, Issue 102). In the United States, recent events in the context of public debates about educational policy in Kansas and Ohio illustrate the growing political influence of proponents of intelligent design.

But what exactly is intelligent design theory? Since the sins of the father are occasionally visited upon the children, it will not go amiss here to begin with an examination of the creation science movement that gave rise to modern intelligent design theory. The first thing worth noting is that while virtually all creation scientists are united in their opposition to secular evolutionary biology (and many are equally repelled by theistic versions of evolution, such as those versions of evolutionary thought that see in evolutionary phenomena the unfolding of God's plan), they disagree among themselves on a wide array of other matters.

Young Earth creationists, for example, maintain that the universe is some 6,000 to 10,000 years old. Modern science, by contrast, estimates the age of the universe at something around fourteen billion years, with the Earth forming some four and a half billion years ago.

Young Earth creationists typically have to reject rather more than just evolutionary biology to fit what we see into their truncated chronology. Vast tracts of modern physics and chemistry, not to mention geology and anthropology, must be largely in error if these theorists are correct. In fact, by seeing the biblical chronology and the events and peoples depicted in the Bible as true and accurate depictions of history, these creationists must also reject many well-established archaeological facts about human history (Davies 1992, 1998; Finkelstein and Silberman 2001; Thompson 1999). In the United States, the Institute for Creation Research (ICR) in California is a leading center for this species of creationism.

While young Earth creationists take the biblical chronology very literally, they are forced to go to fanciful lengths to accommodate modern scientific discoveries. For example, the story of Noah's Ark looms large in many of these religious fantasies, where it is often presented as a genuine zoological rescue mission. In some versions, even the dinosaurs entered the ark two by two. We are told that humans and dinosaurs lived together and that the Grand Canyon was scooped out by a tidal wave during the Great Flood. Mount Ararat, the resting place for Noah's Ark (the Holy Grail sought by numerous creationist expeditions to modern Turkey), is viewed as the source of post-Flood biodiversity, with koala bears presumably following a fortuitous trail of eucalyptus leaves all the way to Australia (then joined, perhaps, to South America, but moving rather quickly ever since). The Jurassic Ark must have been a mighty vessel indeed.

Young Earth creationism, however, has attracted many religious extremists, and it is in this context that one sees the claim developed that evolution is the work of the Devil. Henry Morris of ICR has said of evolution that "the entire monstrous complex was revealed to Nimrod at Babel and perhaps by Satan himself. . . . Satan is the originator of the concept of evolution" (1974, 74–75). And from Nimrod the line of wicked descent presumably runs to Darwin and his contemporary intellectual heirs in the scientific community who refuse to give God, angels, and an assortment of demonic bogeymen a place alongside electrons, quarks, gravitational fields, and DNA in the scientific account of natural phenomena.

Recent investigations have uncovered connections between young Earth creationists at the ICR and Islamic fundamentalists—though after the events of 9/11, these groups would no doubt not like to have this resurface in a public forum. For our purposes, the Turkish experience can be seen as a warning of the dangers that accompany efforts by religious extremists who are bent on the destruction of a secular government. It should serve as an alarm call to those of us in the United States who have so far been silent about the steady erosion of the wall of separation between church and state—a process of erosion that has been accelerated by politicians at local, state, and national levels, who either have their own extreme religious agendas or who have shown themselves to be all too willing to pander to extreme religious voices for the sake of expediency.

Turkish scholars Ümit Sayin and Aykut Kence have noted of the BAV (the Turkish counterpart of the ICR) that:

> BAV has a long history of contact with American creationists, including receiving assistance from ICR. Duane Gish and Henry Morris visited Turkey in 1992, just after the establishment of BAV, and participated in a creationist conference in Istanbul. Morris, the former head of ICR, became well acquainted with Turkish fundamentalists and Islamic sects during his numerous trips to Turkey in search of Noah's Ark. BAV's creationist conferences in April and June 1998 in Istanbul and Ankara, which included many US creationists, developed after Harun Yahya started to publish his anti-evolution books, which were delivered to the public free of charge or given away by daily fundamentalist newspapers. (1999, 25)

Sayin and Kence go on to observe that BAV, though it uses anti-evolution arguments developed by the ICR, has its own unique Islamic objectives; this has been echoed by Taner Edis (1999) in his examination of the relations between ICR and BAV. We should not underplay the significance of these links between ICR and BAV, for Turkey is a major NATO ally.

According to Arthur Shapiro (1999), the links between the ICR and Islamic extremists in Turkey were forged as part of a strategy by extremists in Turkey to undermine the nation's secular government. Shapiro has shown that ICR materials have been adapted to Islamic ends as part of a concerted attack on secular science in particular and secular belief in general. What of ICR's role in all this? Shapiro asks:

> Does ICR care that its Turkish friends are using its materials and assistance to destabilize Turkey? Does it have any concern about the potential effect of political creationism in Turkey on the future of

> NATO or the stability of the Eastern Mediterranean? . . . Its own materials suggest either complete disingenuousness or incredible naïveté. The ICR's *Impact* leaflet number 318, published in December 1999, presents its work in Turkey as an effort to bring the Turks to Christ. But the Turks with whom the ICR is working have little interest in coming to Christ. They are too busy trying to come to power. (1999, p. 16)

Whatever the initial motives were in joining hands with Islamic fundamentalists, it appears that in the hands of Islamic creationists, ICR's anti-Darwinism involves much more than a rejection of secular biological science. It involves a rejection of secular politics and the secular society that supports it.

This last point is supported by an examination of the writings of Islamic creation scientists such as Harun Yahya. Yahya is quite explicit about the alleged connection between Darwinism and secular ideologies as diverse as fascism and communism. In his book, *Evolution Deceit: The Scientific Collapse of Darwinism and Its Ideological Background*, in addition to parroting many fallacious claims about science that appear to descend with little modification from ICR positions (notably absent are ICR claims about the Great Flood), he argues, in curious ecumenical tones, that Darwinism is at the root of religious terrorism, be it done in the name Christianity, Islam, or Judaism:

> For this reason, if some people commit terrorism using the concepts and symbols of Islam, Christianity and Judaism in the name of those religions, you can be sure that those people are not Muslims, Christians or Jews. They are real Social Darwinists. They hide under the cloak of religion, but they are not genuine believers. . . . That is because they are ruthlessly committing a crime that religion forbids, and in such a way as to blacken religion in peoples' eyes.
>
> For this reason the root of terrorism that plagues our planet is not any of the divine religions, but is in atheism, and the expression of atheism in our times: "Darwinism" and "materialism." (2001, 19–20)

While it is hard to credit deception on this scale—even self-deception—the theme is one that will resonate with creationists and other Christian extremists in the United States. That is, religion is never to be assessed in terms of its objective consequences, and secularism (Darwinism in the context of science education) is the root of all evil.

Subtler links to Islam can be found in the context of the intelligent design movement. Muzaffar Iqbal, president of the Center for Islam and Science, has recently endorsed work by intelligent design theorist William Dembski. According to the Web page for the Center for Islam and Science, Islam recognizes the unity of all knowledge: "This is based on the concept of *Tawhid,* Unicity of God, which is the most fundamental principle of Islamic epistemology." The idea that scientific knowledge is unified through knowledge of God is an idea that resonates with intelligent design theorists in the West, who, as we shall see, would like to make it a fundamental principle of Christian epistemology. There is nothing sinister here, save a common interest, crossing religious boundaries, in blurring the distinction between science and religion. Of more concern is the fact that the boundaries to be blurred are boundaries between particular conceptions of science and particular conceptions of religion that both scientists and religious believers may reasonably reject.

Getting closer to home, not all creationists in the West subscribe to young Earth creationism. Thus, old Earth creationists, some through an artful interpretation of the *days* mentioned in Genesis 1 and 2 and some through a genuine respect for the discoveries of modern science, maintain that the Earth is of great antiquity. Old Earth creationists have even welcomed talk of a cosmological big bang, provided that it was an event initiated by God, with subsequent events representing, perhaps, the unfolding of the divine plan. Ideas along these lines can be seen in the writings of some of the cosmological proponents of intelligent design theory, and we will discuss them at length later in the book.

But if these believers in the rock of ages disagree about the age of rocks, it nevertheless remains the case that it is against this background of contradictory views about creation that the modern intelligent design movement manifested itself in the early 1990s. Phillip Johnson, who is the architect of the intelligent design movement, is the intelligent designer of something called the wedge strategy. Johnson (2000a, 13) invites us to imagine that our way is blocked by a large, heavy log. To pass it, we must break it up into pieces. To break it up into pieces, we must find cracks in the log, and drive wedges into these cracks. The wedges will split the log. Natural science is this log that, according to Johnson, is barring our way to Jesus.

Natural science is seen as barring the way to Jesus because it is said to be thoroughly contaminated by a pernicious philosophy known as naturalism. Johnson observes:

> The Wedge of my title is an informal movement of like-minded thinkers in which I have taken a leading role. Our strategy is to drive the thin end of our Wedge into the cracks in the log of naturalism by bringing long-neglected questions to the surface and introducing them to public debate. Of course the initial penetration is not the whole story, because the Wedge can only split the log if it thickens as it penetrates. (2000a, 14)

At the thinnest end of the wedge are questions about Darwinism. As the wedge thickens slightly, issues about the nature of intelligent causation are introduced. As the wedge thickens still further, the interest in intelligent causation evolves into an interest in supernatural intelligent causation. At the fat end of the wedge is a bloated evangelical theology. As Johnson himself observes:

> It is time to set out more fully how the Wedge program fits into the specific Christian gospel (as distinguished from generic theism), and how and where questions of biblical authority enter the picture. As Christians develop a more thorough understanding of these questions, they will begin to see more clearly how ordinary people—specifically people who are not scientists or professional scholars—can more effectively engage the secular world on behalf of the gospel. (2000a, 16)

Reading Johnson's words, I am drawn to think not of woodcutters and their wedges but of the older kids who hang around schoolyards, peddling soft drugs so that a taste for the harder stuff will follow.

For the dark side of the wedge strategy, lurking at the fat end of the wedge, lies in the way that it is intelligently designed to close minds to critical, rational scrutiny of the world we live in. The wedge strategy describes very well the very process whereby, beginning with mild intellectual sedatives, religion becomes the true opiate of the masses. As Johnson makes clear (2000a, 176), once the wedge is driven home, even the rules of reasoning and logic will be have to be adjusted to sit on theological foundations. In this way, critical thinking and opposition will not just be hard but literally unthinkable!

In this book, I am concerned mainly with the issues at the thin end of the wedge, where there are three basic issues. First, there is opposition to the philosophy of naturalism; second (and related to this), there is opposition to evolutionary biology; and, third, there are positive arguments for introducing into science supernatural intelligent causes of natural phenomena. The postulation of such intelligent causes predates the rise of modern science, appearing most notably in the context of medieval Christian theology as the conclusion of an argument for the existence of God, called the *argument from design*. In a way, the thin end of the wedge can be thought of as an expression of the distilled essence of creation science, the veritable wheat minus the chaff, for it is what is left when the silliness about Noah's Ark, global floods, and Fred Flintstone scenarios concerning the coexistence of humans and dinosaurs are scattered to the winds.

Christianity and Creationism

Before I move to consider these issues, I would like to make some observations about science and religion, and Christianity in particular. First, it is false that all Christians are creationists or advocates of creation science. It is false that all Christians are religious extremists. It is also false that all Christians are intelligent design theorists. Indeed, many are deeply offended by such a suggestion. Christianity as we know it today manifests considerable diversity with respect to belief. Creationists and religious fundamentalists most assuredly do not speak for all Christians, though all too often it is the extreme voice of creationists that is heard in public debate.

Importantly, many strands of the diverse cultural fabric of the Christian community have indeed found ways to accommodate science and religion. Such strands include, but are not limited to, Roman Catholics, Episcopalians, Anglicans, Methodists, and Presbyterians. For many Christians, belief in God is about how to go to heaven, and not how the heavens go. In these terms, it is a gross abuse of the Bible, and a truly wretched theology, to think of it as a science primer. And not just Christianity but other religions, too, including Judaism, Islam, and Buddhism, have found ways to have both religion and science and hence to live in the modern world that we all must share, notwithstanding our diverse beliefs.

Phillip Johnson knows this, and he knows that many Christians believe that God works through evolution. Johnson is dismissive. In a reply to criticisms from Cassandra Pinnick and myself, he claimed, "The deep conflict cannot be papered over with superficial solutions such as interpreting the 'days' of Genesis as geological ages or viewing evolution as God's chosen means for bringing about his objectives. . . . God-guided evolution isn't really evolution at all, as scientists use the term; it might better be called *slow creation*" (2000b, 102). He adds: "Sure, you can accept neo-Darwinism and still be "religious"—in a sense. We all know about Dobzhansky, Teilhard, and liberal bishops like John Shelby Spong. But is the theory consistent with the beliefs held by so many that a supernatural being called God brought about our existence for a purpose? That question deserves something better than a cynical evasion" (2000b, 103).

It is true that some adherents of Christianity have indeed a strong propensity to cast the character of their religious beliefs so that they inevitably conflict with science. But science and religion have been coevolving since the events precipitating the rise of modern science took place in the Renaissance. I will relate part of this history in the next chapter. For the present, it is worth noting that there are serious theological alternatives to the religious conservatism that Johnson seems so keen to champion. The advice I gave Johnson—from a good source—back in my review of his work (2000) still seems to be on the mark: first cast out the beam out of thine own eye; and then shalt thou see clearly to cast out the mote out of thy brother's eye.

At this point I must be blunt with you. I am an atheist, and by this I mean that I am someone who does not believe that there is any credible evidence to support belief in the existence of God. By a similar light, I am also an *asantaclausist* and an *aeasterbunnyist*. And I regret to inform you that I have no particular solution to the problem of reconciling science and religion. Sadly, I very much doubt that the problem has a universally acceptable rational solution. Those most in need of such a solution are the very ones incapable of appreciating any such solution, were it to be discovered and offered. We have just seen that the likes of Phillip Johnson have no time for the reasonable Christian folk who have found ways to have their religion and nevertheless accept the results of modern science. You are more likely to reconcile the Israelis and the Palestinians or the Protestants and Catholics in Northern Ireland than you are to come to a universally agreeable solution to the problem of the reconciliation of science and religion.

Nevertheless, it is surely a testimony to the power of science envy in our culture that religious extremists have found it necessary to invent religious versions of science to serve their ends. The supreme irony, of course, is that in passing off their religious views as scientific, intelligent design theorists and creationist fellow travelers seek to ruin the very sciences in whose respectability they try to cloak themselves. The label is appropriated only to be destroyed. Whether we have any reason to take the various proposals for a supernatural science seriously is examined in the course of this book.

The Structure of the Book

In the next chapter, I will examine the argument from design to show where it came from and how it is supposed to work. I will argue that there are two fundamental kinds of design argument. One concerns complex, adapted structures and processes in biology; the other concerns the universe as a whole. Both arguments involve topics about which there are gaps in our current scientific knowledge. I will show how the argument from design, far from being undercut by the rise of modern science, was in fact bolstered by it. I will also discuss some early critical reactions to the argument due, among others, to David Hume and Immanuel Kant. This will provide the backdrop for what follows in the remainder of the book.

In chapter 2, I will examine Darwin's response to the traditional biological version of the argument from design. In addition to examining the details of evolutionary theory, I will also discuss Darwin's attitudes toward religion. This will also be an opportunity to examine developments in evolutionary biology in the 144 years since *The Origin of Species* was first published in 1859. Among the topics discussed will be the impact of genetics on evolutionary biology and recent research bringing together issues in evolution with issues in developmental biology.

In chapter 3, I turn my attention to thermodynamics—partly because errors about the meaning of the Second Law of Thermodynamics pervade creationist literature and partly because the recent study of nonequilibrium thermodynamics has revealed how natural mechanisms, operating in accord with natural laws, can result in the phenomenon of self-organization, whereby physical systems organize themselves into complex, highly ordered states. In addition to evolutionary mechanisms studied by biologists, there are thus other natural sources of ordered complexity operating in the universe. A person ignorant of such mechanisms might well conclude that supernatural causes are in operation where there are in fact none.

Before turning to examine modern design arguments, we need to be clearer about intelligent design theory, its so-called wedge strategy, and what it sees itself as opposing. Supernatural science is thus the subject of chapter 4. One of the central issues to be discussed concerns claims that there are supernatural causes operating in nature to bring about effects beyond the reach of natural causes. Such conclusions, if established, would point to a deficiency in the philosophy of naturalism. Roughly speaking, this is the view that the only legitimate business of science is the explanation of natural phenomena in natural terms; put slightly differently, such causes as there are of natural effects must themselves be natural, as opposed to supernatural. Intelligent design theorists make much of naturalism and its deficiencies. But it is unclear whether the natural sciences, as opposed to particular natural scientists with extrascientific agendas, are actually committed to naturalist philosophy.

Scientists do tend to focus on the search for natural causes for effects of interest, but perhaps this involves less of a prior commitment to a naturalistic philosophy (most scientists in my experience—exceptions duly noted—couldn't give a hoot for philosophy anyway) and is more a reflection of the collective experience of scientists of all stripes over the last 300 years of modern science. We simply have not seen convincing evidence for conclusions supporting the operation of supernatural causes in nature. On this view, while scientists do not categorically reject the possibility of supernatural causation, they do not take it seriously at present either, primarily because of a complete lack of convincing evidence. On this view, the naturalism of the natural sciences may be methodological, reflecting long experience sifting evidence to support causal explanations, rather than philosophical or metaphysical, reflecting intellectual bias ruling out the very possibility of supernatural causation prior to the onset of investigations, the arrival of data, and its subsequent interpretation.

To sharpen these issues, I will examine some recent attempts to introduce supernatural causes into medicine. I refer here to the numerous studies that have been performed and even reported in the scientific literature—in distinguished journals such as *The Archives of Internal Medicine*—that claim empirical support for conclusions about the efficacy of prayer (and related activities such as church-going) as a therapeutic modality. These studies deserve our attention because, independently of whether they are flawed or not, they represent serious attempts to gather evidence in favor of supernatural conclusions (attempts that are simply not in evidence in the intelligent design movement, which has contented itself with extensive armchair theorizing).

In chapter 5, I will present some recent and influential biochemical arguments that have been put forward, by Michael Behe and others, to justify the conclusion of intelligent design. Since biochemistry was essentially an unborn fetus in the body of science in Darwin's day, it is certainly possible that these new arguments are not simply old wine in new bottles but represent a substantial challenge to evolutionary biology. The issue here will hinge on the concept of irreducible complexity, a special type of biological complexity that has been alleged to resist an explanation in evolutionary terms. The biochemical design arguments, as well as their broader implications, will be subject to critical scrutiny. In the course of this analysis, it will be shown how irreducible complexity could have evolved, and some relevant evidence will be discussed.

In chapter 6, I will present arguments for the conclusion of intelligent design that proceed from considerations of the nature of the universe and from anthropic principle cosmology in particular. The cosmological design arguments are shown to be inconclusive. Several problems are identified. In some versions of these arguments, there are errors about causation (especially with respect to thermodynamical reasoning). There are also issues about probability theory and failures to consider relevant, alternative, nonsupernatural hypotheses. There is no good evidence to support the claims of intelligent supernatural design. The lessons learned here about the failings of these arguments ought to serve as guides to the critical analysis of future intelligent design arguments, since these will no doubt be forthcoming as gaps get closed and the theorists of supernatural causation are forced to hop to other, currently empty explanatory niches.

In the concluding chapter, I will end the book with some remarks about science, morality, and God. The intelligent design movement has a social agenda that seems to go well beyond science education.

I will discuss this agenda. Design theorists see the issue of origins as being crucial to the formulation of social, political, and legal policies. At the root of these claims is belief in supernatural causation and an objective, transcendent moral order rooted in God.

By contrast, I believe that Darwin himself provides a way of thinking about the functional role of morality that, when developed, accords well with the democratic values that are our common inheritance from the Enlightenment. At rock bottom, this book is about the Enlightenment and its enemies and about the choices we will all have to make, not just about science, but about life itself: how we want to live, how we want society to be structured, how we want to see the future unfold. Ultimately, it is about what we value and how this reflects differing estimates of the nature of the world we live in.

1

The Evolution of Intelligent Design Arguments

We saw in the introductory chapter that lying at the heart of all species of modern creation science, whether it is the young Earth creationism, old Earth creationism, or intelligent design theory, is the argument from design. This argument has a long evolutionary ancestry (Shanks 2002), with roots trailing back into pre-Christian, heathen philosophy, and in this chapter we will examine the evolution of this centerpiece of contemporary creationist theorizing. The modern design arguments lying at the heart of creation science and its most recent incarnation, intelligent design theory, descend with little modification from a long line of earlier arguments. These arguments belong to an ancient cultural lineage extending back to antiquity and rooted in prescientific speculation about the nature of the universe. Since wine does not necessarily improve with age, and since modern creationist thinking contains much old wine in new designer-label bottles, it will be useful to examine this history in order to appreciate the context in which the modern arguments survive, like tenacious weeds, in the minds of men.

Conceived in Sin: Heathen Origins

To understand the origins of the argument from design, we must go back to pre-Christian ancient Greece. A convenient place to start this magical history tour is with the heathen philosophy and science of Aristotle (384–322 B.C.), teacher of Alexander the Great. Aristotle's ideas will be seen to have a major influence on medieval philosophical theology, especially that of St. Thomas Aquinas (1224–1274), who would give a classic statement of the Christian version of the argument from design.

Aristotle, like many other Greek thinkers of his time, was very interested in the relationship between matter and form. Aristotle contended that in nature we never find matter on its own or form on its own. Everything that exists in nature is a unity of matter and form. This unity of matter and form Aristotle designates as a *substance*. Dogs were one type of substance formed or shaped by the form *dogness*, and mice another, formed or shaped by *mouseness*. Form thus determines species membership. Form is what all members of a species have in common, despite variations in appearance. Species differences reflect a difference with respect to the form shaping matter. Species-determining forms are held to be eternal and changeless, and thus evolution is claimed to be impossible. In this view, the categories of everyday experience are essentially fixed. The study of form is morphology, and Aristotle's thinking on these matters became associated with various morphological species concepts, in which organisms are categorized on the basis of shape. It is not an exaggeration to say that Aristotle's way of thinking has worked much mischief in both science and biology, as we shall see at various points in this book.

To understand what substances are and how they change, Aristotle introduced the idea of the four causes. And since this view of causation will turn out to be of importance later, we must examine the basic details here. The doctrine of the four causes is put forward to explain the changes we see in nature. Of any object, be it an inanimate object, an organism, or a human artifact, we can ask four questions: (1) What is it? (2) What is it made of? (3) By what is it made? (4) For what purpose is it made?

To answer the first question is to specify the *formal cause*, hence to identify substance and species. To answer the second question is to specify the *material cause* and explain the material composition of the object. To answer the third question is to specify the *efficient cause* and explain by what a thing was made or by what a change was brought about. To answer the fourth question is to specify the *final or functional cause*—the function of the object, the end or purpose for which it was made. The view that objects in nature have natural functions or purposes is known as a *teleological* view of nature (from the ancient Greek words *telos*, meaning "purpose," and *logos*, meaning "logic or rational study").

Thus an object might be a mousetrap (formal), made of wood and metal (material), by the mousetrap manufacturer (efficient), to catch mice (functional). But this scheme works for objects that are not human artifacts. An object might be an acorn, made of organic matter, by the parental oak tree, to become an oak tree itself. Importantly for our purposes, Aristotle saw that the form of an object determines its end or function. That is to say, the end or function of an object is determined by its internal nature. This is the sense in which it is the end of an acorn to become an oak tree (Stumpf 1982, 89–92).

For Aristotle, everything in nature, be it organic or inorganic, had a natural end, function, or purpose determined by its form. Yet Aristotle differentiated between organic and inorganic beings through the idea of souls. The soul becomes the form of the living, organized body. An organized body has functional parts, such that when they attain their end, the organized body as a whole is capable of attaining its end. Humans are thus said to have rational souls, and we are defined as rational animals. The parts of the acorn work together that the acorn might become an oak tree; the parts of a human work together that we, too, can achieve our end, which was for Aristotle eudemonia. But what is eudemonia?

The Greek word *eudemonia* is inadequately translated as "happiness," especially as we are apt to understand it today as meaning pleasure, titillation, or even enjoyment. It really means something closer to "well-being" or "general welfare." Nevertheless, eudemonia was seen as the chief human good—the goal, function, or purpose of rational human action. The purpose of human existence, then, is the attainment of this state of well-being. A human is as goal-directed by virtue of its rational nature as the acorn is by its oak tree nature.

In fact, the function of anything in nature can be specified by saying what it is there for the sake of. Aristotle put it this way: "If then we are right in believing that nature makes nothing without some end in view, nothing to no purpose, it must be that nature has made all things specifically for the sake of man" (Sinclair, 1976, 40). But how did nature do this? Aristotle was somewhat vague about this, yet it is clearly an issue that calls out for an explanation of some sort. Perhaps an analogy would help. Human artifacts, after all, serve various functions and are here for the sake of various people. But they are also crafted by artisans with these ends and functions in mind.

Going beyond the works of Aristotle but remaining rooted in ancient Greece, many thinkers saw evidence of design and purpose in nature. Nature's artisan or craftsman was said to be the demiurge (from *demioergós*, meaning "public worker" or "one who plies his craft for the use of the public"). Human artisans and craftsmen eventually came to be differentiated from nature's craftsman by the use of the word *technites* to describe them (*techne*, meaning "artifice or craft," and the modern word *technology*, literally meaning "the rational study of craft or artifice"). The demiurge thus came to be viewed as the maker of the universe. The demiurge of the ancient Greeks was a cosmic craftsman who purposely shapes and models things from preexisting matter. *This hypothetical being was not one who creates something from nothing.* Indeed, the idea that matter is not created but has always existed is an enduring theme in important strands of ancient Greek thought. The demiurge is thus a shaper of preexisting stuff, not a creator of stuff from nothing.

By the time these heathen intellectual traditions had reached the Roman commentator Cicero, two distinct strands of reasoning about design-with-purpose had appeared—a cosmological strand and a biological strand. Cicero explained the cosmological strand of designer reasoning as follows:

> Again, the revolutions of the sun and moon and other heavenly bodies, although contributing to the maintenance of the structure of the world, nevertheless also afford a spectacle for man to behold . . . for by measuring the courses of the stars we know when the seasons will come round. . . . And if these things are known to men alone, they must be judged to have been created for the sake of men. (Rackham 1979, 273)

In a similar vein, the biological strand of designer reasoning was explained as follows: "Then the earth, teeming with grain and vegetables of various kinds, which she pours forth in lavish abundance. . . . Men do not store up corn for the sake of mice and ants but for their wives and children and households. . . . It must therefore be admitted that all this abundance was provided for the sake of men" (Rackham 1979, 274–275). As the argument from design evolved, two distinct strains emerged—a celestial strain and a terrestrial strain—and both strains, moving from the minds of heathens to pastures new, found ways to invade the minds of Christians.

Roots of Christian Designer Theology

According to the *Catholic Encyclopedia* (www.newadvent.org), the concept of the demiurge also played a role in the thought of early Gnostic Christians, who "conceived the relation of the demiurge to the supreme God as one of actual antagonism, and the demiurge became the personification of the power of evil, the Satan of Gnosticism, with whom the faithful had to wage war to the end that they might be pleasing to the Good God." But while the idea of a demiurge took this turn in Gnostic hands, the idea of a cosmic craftsman would reappear in medieval Christian thought, duly clad in godly trappings.

To see what happened then, first we need to look at the concepts of *potentiality* and *actuality* in Aristotle's thought, for there is the seed of an idea here that will mutate and flower in some interesting ways in the medieval thought of St. Thomas Aquinas. The oak tree giving rise to the acorn is an actual tree; the acorn is a potential oak tree. From this, Aristotle observes that for a potential thing (an acorn) to become an actual thing, there must be a prior actual thing (the parental oak tree). To explain how there can be a world containing potential things that can become actual things, Aristotle thought that there must be a being that was pure actuality, without any potentiality. Such a being would be a precondition for the existence of potential beings that can be subsequently actualized. Aristotle called this being the Unmoved Mover. Exactly what sort of a being Aristotle was trying to talk about is a little vague. But it was not so for St. Thomas Aquinas.

For Aquinas, the unmoved mover was the Christian God. In many ways, Aquinas can be thought of as having made Aristotle's heathen philosophy safe for Christians. Aquinas offered five "proofs" for the existence of God, one of which mirrors the pattern of reasoning that led Aristotle to postulate an unmoved mover. But another of Aquinas's proofs is much more important for our present concerns. It is the celebrated argument from design, an argument that in various mutated forms has worked much mischief on human thinking about the natural world. While Darwin's theory of evolution can be viewed as a sustained refutation of the argument from design as it would descend and evolve with little modification into the niche afforded by natural theology in the eighteenth century, the argument, as we shall see in later chapters, has been resurrected in the writings of contemporary creation scientists and by intelligent design theorists in particular.

The fifth way that Aquinas tried to prove the existence of God—an argument that was intended to be persuasive to rational atheists, who might then heed its message—goes as follows:

> We see how some things, like natural bodies, work for an end even though they have no knowledge. The fact that they nearly always operate in the same way, and so as to achieve the maximum good, makes this obvious, and shows that they attain their end by design, not by chance. Things that have no knowledge tend towards an end only through the agency of something which knows and also understands, as an arrow through an archer. There is therefore an intelligent being by whom all natural things are directed to their end. This we call God. (Fairweather 1954, 56)

In the natural world around us, we observe all manner of seemingly purposeful regularities in the behavior of things that do not possess intelligence. For example, there are regularities in the motions of the tides and in the motions of heavenly bodies—they appear to move in a purposeful manner. Bees make honey, cows make milk, and thus they seem to have a place and a purpose in nature's economy. The behavior of body parts, such as eyes, hearts, and lungs, seems also to be purposive and functional. By analogy with functional artifacts made by human craftspeople that achieve their functions as the result of deliberate design manifesting various degrees of intelligence, nature's artifacts must also have an intelligent designer, one vastly more intelligent than any merely human artisan. And thus, into the yawning gaps in medieval knowledge of the natural mechanisms that give rise to observable phenomena, God-the-designer found a large, cozy niche.

God as Cosmic Engineer

Important as an understanding of medieval philosophical theology is for the purposes of this book, it is also important to give due consideration to medieval technology. The way people interact with the world, through the crafts they practice, the skills they possess (and observe in others), and the machines they make to achieve their own ends and goals, provides the intellectual background, tools, metaphors, analogies, and associated imagery whereby people come to terms with the world around them. We find it very natural to conceptualize that which is strange, alien, and puzzling by the use of metaphors and analogies that are drawn from more familiar domains of human experience and activity. This is especially true when those experiences and activities have yielded fruit of great value to us.

For example, today we can see the broader cultural influences of computer technologies. We do not have to look far to find people trying to make sense of the difference between mind and body by using the computational metaphors of software and hardware; others talk about genetic codes and programs and about genetically programmed behaviors and ways of thinking. But though computers can simulate many interesting phenomena, sometimes the real divide between the computational metaphor and its puzzling subject is not as clearly drawn as we might like. Disputes about these matters arise, for example, in the context of debates about artificial intelligence. The problem is that metaphors are seductive precisely because they enable us to get a handle on the unfamiliar. They can bewitch us, and many before and since the time of Aquinas have been trapped in ways of thinking prompted by the very analogies and metaphors they used to comprehend that which was initially puzzling.

It is very easy—and often very misleading—to move from the claim that something puzzling that has caught our interest appears to us *as if* it is like something else we are familiar with, to the very different claim that this puzzling thing *is literally* like this familiar thing in crucial respects (perhaps even identical). Thus it is one thing to say that in certain circumstances the mind behaves as if it is a computer and quite another, with a very different evidential burden, to say that it literally is a computer. The latter is a much stronger claim than the former, and whereas the former statement may be a useful heuristic claim, the latter may turn out to be quite false and very misleading. As noted before, these matters are debated extensively by folk in the artificial intelligence community. We do not need to settle the dispute one way or the other to appreciate its importance.

Another example may help here. At the beginning of the twentieth century, physicists were struggling to come to terms with the relationship between electrons (just discovered by J. J. Thomson in 1897) and the nuclei or "cores" of atoms. Some people thought electrons were embedded in the nucleus, as if they were like raisins in a bran muffin. But the atom-as-muffin model (actually their gustatory analogy was that of the plum pudding) had to be abandoned: Electrons, unlike raisins, repel each other through electrostatic forces. The as if clause, though undoubtedly helpful early on in these inquiries, did not translate into an is literally clause. Ernest Rutherford suggested a planetary model in terms of which the electrons orbited the nucleus like planets orbiting a sun. This was once again fruitful, since it suggested that it might be important to examine the shapes of electron orbits and their orbital velocities. But the is literally clause was not forthcoming, because according to physics as it was then understood (Maxwell's equations, in particular) electrons orbiting a nucleus should radiate electromagnetic energy, thus spiraling into the nucleus as they lost energy in this way. If the model was right, matter should have collapsed long ago. This puzzle was ultimately resolved in the quantum theory, but in the process we learned that electrons are nothing like macroscopic objects such as planets (or even baseballs or bullets) and that they obey very different rules.

These remarks are relevant here because medieval society in Europe was a mechanically sophisticated society. While the coupling and subsequent coevolution of science and technology that was to accompany the rise of modern science had not yet happened, this should not blind us to the broader cultural importance of machines and machine making in medieval society (Shanks 2002). Today, the visible remnants of medieval society are primarily churches, cathedrals, and castles. Their mechanical accomplishments, often made of wood and leather, have all but perished. Yet those that have survived, along with extensive writings and drawings, testify to a society fascinated by machinery and its possibilities.

In the late medieval period, before the rise of modern science, clock-making skills and the mechanical fruits of those skills had begun to provide useful analogies and comparisons to those concerned with the systematic study of nature. Thus, Rossum remarks: "Parisian natural philosophy at the end of the fourteenth century honored clockmakers by comparing the cosmos or creatures with artful clockworks and the creator-God with a clockmaker. As constructors who designed and built their products, clockmakers thus took their place alongside architects, who were highlighted in these comparisons" (1996, 174). Mechanical artifacts such as clocks provided important metaphors in the struggle to understand the nature of nature. And they helped to crystallize a mechanical picture of nature in which there was purposeful, intelligent design on a cosmic scale. These metaphors were crucially important for an understanding of the purposes served by organisms and the functions of the parts of those organisms.

Organisms as Machines

Modern science as we know it today results from a series of cultural and intellectual changes that occurred in the sixteenth and seventeenth centuries, and these events were profoundly influenced by the medieval experience with machine technologies.

As a medical interest in anatomy and physiology began to germinate and blossom in the early Renaissance, investigators began to conduct systematic inquiries, first into the structure (anatomy) of bodies and then into the functions (physiology) of the parts that the wholes may achieve their appointed purposes. These studies required extensive dissection of dead humans and animals and also vivisection of live animals. These developments in early science played an important role in the evolution of the argument from design.

Some readers of this book may recall dissecting dead rats or frogs in school. Some readers may recall butchering animals for food (or watching others do it). A much smaller number of readers will have dissected human cadavers or performed surgery on live humans or other animals. And anyone with any of this sort of experience will almost certainly have had the benefit of knowledgeable teachers and reasonably accurate textbooks. This was not so for many of the pioneering investigators of the Renaissance, whose teachers may never have dirtied their hands in the practice of dissection or vivisection, leaving such grim work to illiterate assistants, while they read to their students from highly unreliable anatomical "authorities."

Andreas Vesalius (1514–1564) was perhaps the greatest of the Renaissance anatomists, and his book, *The Structure of the Human Body*, was published in 1543—the same year that saw the publication of Copernicus's *Of the Rotation of Celestial Bodies*. Vesalius deserves attention partly because he corrected errors in earlier anatomical traditions—he showed that men and women had the same number of ribs, contrary to biblical authority (I still catch students out with this one)—but partly because he emphasized the importance of direct experimental observation, rather than blind reliance on authority.

Pioneers like Vesalius had to go into this anatomical territory alone, groping their way along with little by way of accurate maps and guides. On entering unknown territory, it was very natural for them to draw on metaphors and analogies derived from more familiar and settled aspects of their experience. The metaphors they drew on were suggestive and helpful in coming to terms with this new and alien experience of the insides of animals. Thus, part of the explanation for the blossoming of anatomical and physiological inquiry lies in the way that Renaissance investigators became increasingly reliant on mechanical metaphors to conceptualize the objects of their inquiries—bodies—in mechanical terms. The metaphor of body-as-machine evolved from crude mechanical analogies (e.g., lungs as bellows) early in the Renaissance to a fully crystallized and articulated mechanical picture of human and nonhuman animal bodies by the middle of the seventeenth century.

The metaphor of body-as-machine had enormous implications for medical inquiries. But we will also see that the mechanical metaphors that fueled the growth of anatomical and physiological inquiry also had broader implications, helping to reinforce the idea of nature-as-machine. It is arguably no accident that a method that had proved so fruitful for physicians should come to shape early inquiries by physicists as well. Somewhere in this process our intellectual ancestors made a transition from seeing nature *as if* it was a machine, with many and complex mechanical components, to seeing it *literally as* a machine, with sundry mechanical wheels within wheels. And to anticipate the relevance of this intellectual transition, real machines need designers and makers. God, as the intelligent designer of the natural machine, was just one of the ways in which early modern science and religion came to enjoy a cooperative relationship—a relationship that would be soured only by events, forced in large measure (but by no means exclusively) by a growing understanding of the consequences of Darwin's theory of evolution.

Medicine and the Rise of Machine Thinking

The role of machine thinking is very clear in the writings of the seventeenth-century anatomist and physiologist William Harvey (1578–1657). Harvey's crucial use of mechanical metaphors can be found in the context of work on the motions of the heart—published in 1628 as *Of the Motions of the Heart and Blood.* The problem confronting Harvey was understanding the complex motions of the heart. Here was a gap in our knowledge that needed filling. And as Harvey himself notes, "I was almost tempted to think with Fracastorius, that the motion of the heart was only to be comprehended by God" (Clendening 1960, 155).

The problem was generated by the speed with which the heart's motions occur, especially in mammals whose hearts had been exposed to public view without benefit of anesthesia and who consequently were in great physical distress. Harvey needed subjects in which the motions of the heart were slower so that the component motions could be resolved. Cold-blooded creatures were most useful in these inquiries, and frogs in particular were very useful, as their hearts will continue to beat a short while after they have been excised from the body. Not for nothing was the frog known as the Job of physiology!

Harvey analyzed the complex cardiac motion into component motions associated with structures discernible in the heart (*ventricles* and *auricles*, the latter being the old word for *atria*). Harvey was then able to synthesize his understanding of the properties of the parts, and their mutual relationships, into a unified understanding of the complex motion of the whole system:

> These two motions, one of the ventricles, another of the auricles, take place consecutively, but in such a manner that there is a kind of harmony or rhythm preserved between them, the two concurring in such a wise that but one motion is apparent. . . . *Nor is this for any other reason than it is in a piece of machinery, in which, though one wheel gives motion to another, yet all the wheels seem to move simultaneously*; or in that mechanical contrivance which is adapted to firearms, where the

> trigger being touched, down comes the flint, strikes against the steel, elicits a spark, which falling among the powder, it is ignited, upon which the flame extends, enters the barrel, causes the explosion, propels the ball, and the mark is attained—all of which incidents, by reason of the celerity with which they happen, seem to take place in the twinkling of an eye. (Clendening 1960, 161, my italics)

In this passage, we see how the explicit use of mechanical metaphors could yield natural resolutions of problems that had hitherto been viewed as mysteries beyond the reach of human ken.

Thinking of the operation of the heart in mechanical terms—and hence as a system admitting of a quantitative description—yielded further fruits. Even granting a large margin of error, Harvey estimated that in an hour the heart could pump more blood than the weight of its human owner. Where was all this blood coming from, and where did it go? Harvey had a radical solution. There is a mystery.

> Unless the blood should somehow find its way from the arteries into the veins, and so return to the right side of the heart; I began to think whether there might not be A MOTION, AS IT WERE, IN A CIRCLE. Now this I afterwards found to be true; and I finally saw that the blood, forced by the action of the left ventricle into the arteries, was distributed to the body at large, and its several parts, in the same manner as it is sent through the lungs, impelled by the right ventricle into the pulmonary artery, and it then passed through the veins and along the vena cava, and so round to the left ventricle in the manner already indicated. (Clendening 1960, 164)

Harvey thereby united his own research on the structure and function of the heart with earlier work on pulmonary circulation to conceptualize the conjoined system of heart and blood vessels as a closed, mechanical circulatory system. But even as machine thinking closed these gaps in our knowledge, it should be obvious that the very employment of machine metaphors invited theological speculation.

Surveying these events, it is fair to say that correlative with the rise of modern science is the dual phenomenon of nature being conceptualized with the aid of mechanical metaphors and nature being studied with the aid of machines (telescopes, microscopes, barometers, vacuum pumps, and so on). It was the incredible success of this new way of thinking and this new way of exploring nature that cemented the union between science and technology—a union that owes its existence in no small measure to the work of investigators in anatomy and physiology.

More important, in the course of the seventeenth century, nature itself came to be seen as a complex system of interacting bodies in motion that could be understood in mechanical terms. Arguably, the crowning achievement of seventeenth-century physics is to be found in Sir Isaac Newton's (1642–1727) great work, *Mathematical Principles of Natural Philosophy*, published in 1687. The resulting system of physics—Newtonian mechanics—provides a vision of the universe itself as a giant machine whose parts are held together, and whose motions are interrelated, through gravitational forces.

In Newton's England, the emergence of modern science in the seventeenth century started to initiate cultural changes, especially with respect to science and its relationship to religion, as witnessed by John Aubrey:

> Till about the yeare 1649 when Experimental Philosophy was first cultivated by a Club at Oxford, 'twas held a strange presumption for a Man to attempt an Innovation in Learning; and not to be good Manners, to be more knowing than his Neighbours and Forefathers; even to attempt an improvement in Husbandry (though it succeeded with profit) was look'd upon with an ill Eie. Their Neighbours did scorne to follow it, though not to doe it, was to their own Detriment. 'Twas held a Sin to make a Scrutinie into the Waies of Nature; Whereas it is certainly a profound part of Religion to glorify God in his Workes: and to take no notice at all of what is dayly offered before our Eyes is grosse Stupidity. (Dick 1978, 50–51)

Though atheism was almost unthinkable in Aubrey's day, scientific scrutiny into the ways of nature would indeed lead investigators to question whether the works before them were the works of God or the fruits of the operation of natural mechanisms in accord with the scientific laws of nature. And a horror of new ideas, especially the fruits of scientific inquiry, and a reluctance to "rise above your raising" were evidently as prevalent among Aubrey's contemporaries as they are among religious fundamentalists today.

The Intelligent Design of the World

The mechanical picture of the universe that crystallized and came to fruition in seventeenth-century science contained a vision of organisms as nature's machines—machines that seemed to fit into the world in which they were found. Each seemed to have a natural place in the economy of nature. Each was clearly adapted to a place in the environment. As for further observations of the adapted nature of animal behavior—for example, the nest building of birds and the return of swallows in the spring—as well as observations of physiological, morphological, and anatomical adaptation, these were evidences of providential machine design. For the scientist at the end of the seventeenth century, these features of the organic world were captured by the title of John Ray's (1627–1705) book, *The Wisdom of God manifested in the Works of Creation* (1693).

The picture of organisms that emerged from seventeenth-century science is filled with mechanical metaphors: stomach as retort, veins and arteries as hydraulic tubes, the heart as pump, the viscera as sieves, lungs as bellows, muscles and bones as a system of cords, struts, and pulleys (Crombie 1959, 243–244). The metaphors bolster a picture of organisms as special machines made by God. As the philosopher Leibniz put it in the *Monadology* (1714):

> Thus each organic body of a living thing is a kind of divine machine, or natural automaton, which infinitely surpasses all artificial automata. Because a machine which is made by the art of man is not a machine in each of its parts; for example, the tooth of a metal wheel has parts or fragments which as far as we are concerned are not artificial and which have about them nothing of the character of a machine, in relation to the use for which the wheel was intended. But the machines of nature, that is to say living bodies, are still machines in the least of their parts *ad infinitum*. This it is which makes the difference between nature and art, that is to say, between Divine art and ours. (Parkinson 1977, 189)

Thus, organisms, unlike watches, are machines all the way down, and this is what differentiates God's handicraft from that of mere mortal mechanics.

But inorganic nature, too, was seen in mechanical terms. As noted previously, Newton's universe is a clockwork universe—a giant machine with many interacting, moving parts. And wheels within wheels could be seen everywhere. Not only did the organism have its mechanical parts each adapted for specific functions necessary for life but also different organisms had distinct places in nature. Specialized in distinct and unique ways, they, like the parts within them, had proper places in the natural machine.

The intellectual tradition of studying nature—the mechanical fruit of God's providential design—in order to make discoveries about the creator (both his very existence, as well as particular properties, such as benevolence) is known as *natural theology*. We have already seen that a version of the argument from design was formulated in the medieval period. But the argument, far from being dispelled by the rise of modern science, was in fact bolstered by it. Prior to Darwin, natural science and natural theology were coupled enterprises, with figures prominent in one of these intellectual enterprises often being prominent in the other. This was particularly true of Sir Isaac Newton.

Newton and Design in Nature

The two main lines of modern reasoning about intelligent design—design of the universe as a whole (cosmological design) and design of organisms (biological design)—are present in Newton's writings on natural theology. We are all creatures of our times, and Newton was no exception. Newton's times were times when scientists could seriously entertain natural theology, just as the times of St. Thomas Aquinas were times when it was intellectually respectable to entertain the ideas that the Earth was at the center of the universe, that there were but four elements, and that infectious disease was caused by sin.

It was arguably no accident that Newton, the father of classical mechanics in physics, should have articulated a version of the cosmological design argument in the context of natural theology; after all, he was an heir to a rich inheritance of mechanical thinking that had been intertwined with theological speculation. As Newton himself put it:

> The six primary planets are revolved about the sun in circles concentric with the sun, and with motions directed toward the same parts and almost in the same plane . . . but it is not to be conceived that mere mechanical causes could give birth to so many regular motions, since the comets range over all parts of the heavens in very eccentric orbits. . . . This most beautiful system of the sun, planets and comets could only proceed from the counsel and dominion of an intelligent and powerful Being . . . and lest the systems of the fixed stars should, by their gravity, fall on each other, he hath placed those systems at immense distances from one another. (Thayer 1953, 53)

Like Aquinas before him, Newton was impressed with the natural motions observed in the heavens and saw in them evidence of providential design.

Importantly for the present purposes, Newton saw evidence of intelligent design in the biological world, too:

> Opposite to godliness is atheism in profession and idolatry in practice. Atheism is so senseless and odious to mankind that it never had many professors. Can it be by accident that all birds, beasts and men have their right side and left side alike shaped (except in their bowels); and just two eyes, and no more, on either side of the face . . . and either two forelegs or two wings or two arms on the shoulders, and two legs on the hips, and no more? Whence arises this uniformity in all their outward shapes but from the counsel and contrivance of an Author? (Thayer 1953, 65)

For Newton, morphological similarities were evidence of deliberate intelligent design. Atheism was odious because it could offer no good account of the similarities, save that they were, perhaps, fortuitous accidents.

But Newton does not rest his case simply with the observation of morphological similarities. There is also evidence of adapted complexity:

> Whence is it that the eyes of all sorts of living creatures are transparent to the very bottom, and the only transparent members in the body, having on the outside a hard transparent skin and within transparent humors, with a crystalline lens in the middle and a pupil before the lens, all of them so finely shaped and fitted for vision that no artist can mend them? Did blind chance know that there was light and what was its refraction, and fit the eyes of all creatures after the most curious manner to make use of it? These and suchlike considerations always have and ever will prevail with mankind to believe that there is a Being who made all things and has all things in his power, and who is therefore to be feared. (Thayer 1953, 65–66)

For Newton, such adapted complexity had two possible explanations: first, that it was the result of intelligent design or, second, that it all came about by chance and happenstance. Newton is inclined to the former, as the latter is—and everyone will admit this—so implausible as to be silly and beyond belief. Part of Darwin's achievement, as we shall see, is to offer a third possibility—one that Newton never considered—to explain the same appearances in nature. Darwin's views will be examined in the next chapter.

Newton, though clearly a believer in both God and creation, was no biblical literalist, and this sets him apart from many contemporary advocates of creation science. As Newton himself put it in a letter to Thomas Burnet, "As to Moses, I do not think his description of the creation either philosophical or feigned, but that he described realities in a language artificially adapted to the sense of the vulgar" (Thayer 1953, 60), adding:

> If it be said that the expression of making and setting two great lights in the firmament is more poetical than natural, so also are some other expressions of Moses, as when he tells us the windows or floodgates of heaven were opened.... For Moses, accommodating his words to the gross conceptions of the vulgar, describes things much after the manner as one of the vulgar would have been inclined to do had he lived and seen the whole series of what Moses describes. (Thayer 1953, 63–64)

Contemporary biblical literalists and young Earth creationists manifest what Newton called "the gross conceptions of the vulgar." By refusing to accommodate itself to these conceptions, the modern intelligent design movement is intellectually closer to natural theology as Newton understood it. For Newton, there is no conflict between science and religion, and his own account of nature, especially organic nature, was thoroughly intertwined with his religious beliefs.

Paley and the Evidences of Design

William Paley's (1743–1805) great work, *Natural Theology, or Evidence of the Existence and Attributes of the Deity, collected from the Appearances of Nature*, was published in 1802. It was a book that Darwin read and admired. Modern biological creation science and intelligent design theory descend with little modification from the positions articulated by Paley. Paley did give some consideration to astronomy, but observed that "astronomy is not the best medium through which to prove . . . an intelligent creator, but that, this being proved, it shows beyond all other sciences the magnificence of his operations" (quoted in Rees 2001, 163). In chapter 6, I will examine arguments for intelligent design rooted in astronomy and cosmology. The focus here is on Paley's biological arguments.

Like earlier natural theologians, Paley is impressed by his observations of the way organisms show adaptation to their natural surroundings. Organisms contain structures serving specific functions that enable them to fit into their allotted places in nature. In the grand tradition of thinking in terms of mechanical metaphors and analogies, Paley reasons as follows:

> In crossing a heath, suppose I pitched my foot against a *stone*, and were asked how the stone came to be there, I might possibly answer, that for anything I knew to the contrary it had lain there for ever... But suppose I had found a *watch* upon the ground... I should hardly think of the answer which I had before given, that for anything I knew the watch might always have been there. (1850, 1)

Watches are machines with many finely crafted, moving parts adjusted so as to produce motions enabling the whole device to keep time. It would make sense to infer, in the case of such a functional piece of complex machinery, that "we think it inevitable, that the watch must have had a maker—that there must have existed at some place or other, an artificer or artificers who formed it for the purpose which we find it actually to answer, who comprehended its construction and designed its use" (1850, 10). The next step in the argument is to consider the eye, which, like the watch, appears to be a complex piece of machinery with many finely crafted, moving parts, all enabling the organ to achieve its function.

Eyes are compared to telescopes, and Paley is led to the conclusion that the eye, like the watch and the telescope, must have had a designer (1850, ch. 3). More than this, Paley compares the eyes of birds and fishes and concludes, "But this, though much, is not the whole: by different species of animals, the faculty we are describing is possessed in degrees suited to the different range of vision which their mode of life and of procuring their food requires" (1850, 27). Different species occupy different places in nature, and for each species, the machinery of the eye has been fashioned to suit the needs consequent upon their allotted place. Nature thus contains many wheels, and wheels within wheels, all standing as evidence of a mighty feat of engineering and design.

In his discussion of the fruits of comparative anatomy, Paley explains these similarities and differences with the aid of mechanical metaphors:

> Arkwright's mill was invented for the spinning of cotton. We see it employed for the spinning of wool, flax, and hemp, with such modifications of the original principle, such variety in the same plan, as the texture of those different materials rendered necessary. Of the machine's being put together with design . . . we could not refuse any longer our assent to the proposition, "that intelligence . . . had been employed, as well in the primitive plan as in the several changes and accommodations which it is made to undergo." (1850, 143)

Comparative anatomy, then, yields, as it did for Newton, further evidence of intelligent design in the natural world, with mechanical metaphors carrying much explanatory weight.

Could chance or natural causes be behind the adapted complexity we see in nature? Paley was uncompromising on this topic: "In the human body, for instance, *chance, that is, the operation of causes without design*, may produce a wen, a wart, a mole, a pimple, but never an eye. . . . In no assignable instance has such a thing existed without intention somewhere" (1850, 49, my italics). Notice that Paley equates chance not with uncaused events but with events that may have natural causes but that are unguided by intelligence. For the present, what explanations could there be of such complex, adapted structures than deliberate design?

In Paley's day, nearly sixty years before the publication of Darwin's *Origin of Species*, there had been speculation about the possibilities for evolution. And it is clear that he had some acquaintance with naturalistic, evolutionary hypotheses, however fanciful they may have been, that attempted to explain the appearance of adapted complexity without the existence of a supernatural designer.

Paley, as Gould (1993, ch. 9) has noted, had sufficient courage of his convictions that he was prepared to seriously consider alternatives to his proposed scheme of intelligent design. Among these alternatives are evolutionary alternatives:

> There is another answer which has the same effect as the resolving of things into chance; which answer would persuade us to believe that the eye, the animal to which it belongs, every other animal, every plant, indeed every organized body which we can see, are only so

> many out of the possible varieties and combinations of being which the lapse of infinite ages has brought into existence; that the present world is the relic of that variety; millions of other bodily forms and other species having perished, being, by the defect of their constitution, incapable of preservation, or to continuance by generation. (1850, 49–50)

In this passage, we see a role for variation and for differential reproductive success. Darwin, who had studied Paley carefully, must have noticed this passage. But Paley did not see how to develop the ideas, and in the same discussion, the insights are lost.

Paley loses evolutionary insights for at least three reasons: First, he had no real appreciation for the extent of the extinction of earlier species, owing, no doubt, to the fact that the science of paleontology in his day was essentially an unborn fetus, and the idea of extinction was as much an offense to God's plan as was the origination of new species:

> We may modify any one species many different ways, all consistent with life, and with the actions necessary to preservation. . . . And if we carry these modifications through the different species which are known to subsist, their number would be incalculable. No reason can be given why, if these deperdits ever existed, they have now disappeared. Yet if all possible existences have been tried, they must have formed part of the catalogue. (1850, 50)

Second, he had no mechanism to drive the process he describes. The third reason that Paley missed the evolutionary insight had to do with the state of systematics in his day, which, unlike modern, evolutionary approaches to systematics, had no historical component (because none was deemed necessary):

> The hypothesis teaches, that every possible variety of being hath, at one time or another, found its way into existence—*by what cause or in what manner is not said*—and that those which were badly formed perished; but how or why those which survived should be cast, as we see the plants and animals are cast, into regular classes, the hypothesis does not explain; or rather the hypothesis is inconsistent with this phenomenon. (1850, 51, my italics)

For Paley, regularity in the form of the taxonomic order seen in nature (the division of organic beings into plants and animals and subdivisions of each into genera, species, and subspecies) is not a convenience imposed by systematists—"an arbitrary act of mind"(1850,

51)—but reflects an underlying intentional order and plan. That the observable taxonomic order might reflect the operation of evolutionary mechanisms involving descent from common ancestors with subsequent evolutionary modification, over long periods of time, thereby being neither the result of intelligent design nor the mere caprice of systematists, is not considered.

Undergirding Paley's grand scheme of argument is his intellectual inheritance of the conception of nature-as-machine composed in part of organisms-as-machines. Paley, far from bucking the science of his day, was entirely consistent with it:

> What should we think of a man who, because we had never ourselves seen watches, telescopes, stocking-mills, steam-engines, etc., made, knew not how they were made, nor could prove by testimony when they were made, or by whom, would have us believe that these machines, instead of deriving their curious structures from the thought and design of their inventors and contrivers, in truth derive them from no other origin than this, namely, that a mass of metals and other materials having run, when melted, into all possible figures, and combined themselves into all possible forms. . . . These things which we see are what were left from the incident, as best worth preserving, and as such are become the remaining stock of a magazine which, at one time or other, has by this means contained every mechanism, useful and useless, convenient and inconvenient, into which such like materials could be thrown? (1850, 51)

But the possibility remains that organisms are not like machines at all and, if so, that the processes by which they originate and change are nothing like the fruits of intentional design and engineering processes. If organisms are not machines, it is no longer absurd to deny design. But that will involve a scientific revolution in the truest sense.

In the next chapter, I turn to examine Darwin's theory of evolution. There it will be seen that Darwin, in getting away from the idea that organisms are deliberately designed machines, fitting their niches like cogs in nature's grand mechanism, saw a need for a radical reappraisal of what we are and how we stand in relation to other organisms. Darwin's response to Paley is, in fact, a response to a whole way of thinking about organic nature that goes back to the origins of modern science itself. In a way, his work is far more revolutionary than that of Newton, for whereas Newton is a champion for a pre-existing mechanical tradition, Darwin is the initiator of a radical new way of viewing organic nature. But our survey of the argument from design is not quite done, and even before Darwin's meteor struck the world of ideas, concerns about what the argument from design could and could not show had become apparent.

The Age of Reason and the Argument from Design

The eighteenth century, the age of Enlightenment, saw the dawn of the industrial revolution; the spread of technologies rooted in coal, iron, and steam; and the beginning of the social changes that, continued in the nineteenth century, would culminate in the modern, urbanized, industrial economies of the twentieth century. It was also the time of the American Revolution in 1776, the French Revolution in 1789, and the gradual emergence and spread of secular, democratic ideals in politics. Importantly, it was the time of David Hume (1711–1776) and Immanuel Kant (1724–1804), two of the great philosophers this period produced. Both raised concerns about the argument from design. Kant's concerns, though very serious for Christian apologists, were less far-reaching, and I will discuss his first.

Kant's analysis of the argument from design can be found in his *Critique of Pure Reason*, published in 1781. Kant is respectful of the argument from design, for of all the arguments for the existence of God, "it is the oldest, the clearest, and that most in conformity with the common reason of humanity. It animates the study of nature, as it itself derives its existence and draws new strength from that source" (Meiklejohn 1969, 363). Given the way in which Michael Behe, a leading light of the contemporary intelligent design movement, has recently taken the argument from design out of the context of organic anatomy and recast it in terms of the anatomy of biochemical pathways, it is hard to argue with Kant on this point.

As Kant points out, human artifacts result from the intelligence of craftsmen who cause these objects to exist by forcing or causing nature to bend to their wills. They do this by literally reshaping, rearranging and re-forming the stuff of nature. The argument from design requires that the same type of causality involving understanding and will, this time of a supreme intelligence, be operative in the causation of the shapes and forms of things in general, including organisms and even the universe that contains them. Put this way, it is clear that the argument rests both on an analogy between nature and the products of the craftsman and upon the notions of understanding and will as causal factors in the production of artifacts. I note here, and I will return to this point in a later chapter, that without an analysis of these concepts that displays their causal role very clearly, an appeal to them as causal factors in the production of anything, let alone universes and organisms, will be little better than the stage magician's appeal to the magic word *abracadabra* in the production of a rabbit from a hat!

But this is to look ahead, and for the present, I notice that Kant's worry is a different one, for as he observed of the argument from design:

> The connection and harmony existing in the world evidence the contingency of the form merely, but not of the matter, that is, of the substance of the world. To establish the truth of the latter opinion, it would be necessary to prove that all things would be in themselves incapable of this harmony and order, unless they were, even as regards their *substance*, the product of supreme wisdom. But this would require very different grounds of proof than those presented by the analogy with human art. The proof can at most, therefore, demonstrate the existence of an architect of the world, whose efforts are limited by the capabilities of the material with which he works, but not of a creator of the world, to whom all things are subject. (Meiklejohn 1969, 364–365)

Thus, if the argument from design works, it supports at most the existence of a cosmic craftsman or engineer who, like a human craftsman or engineer, imposes his will and understanding on pre-existing matter and whose creative capabilities are limited by the properties and dispositions of that matter. A bad workman may blame his tools, but even a skilled craftsman cannot get something from nothing and is limited in his works by the materials he deals with.

The argument from design thus does not support the existence of a creator who first has the causal power to make something from nothing—a feat required by the God of Christianity—so that he can fashion the materials so produced. The argument simply will not support ambitious Christian conclusions, and for all the massage and manipulation, the cosmic craftsman of the argument from design is hardly different from the demiurge of heathenism from which it was derived. Christian apologists need not a designer who is not a creator, or a creator who is not a designer, but a designer-creator. Kant's point is that the argument from design points only toward a designer. It does not justify the other half of God's supposed nature.

By contrast, Hume is more concerned with the issue of the inference to design itself, as it appears in the argument from design. Before turning to this issue, I would like to draw your attention to a passage in Hume's *Dialogues Concerning Natural Religion* (published after his death, in 1779), where he makes the following observations:

> For ought we can know *a priori*, matter may contain the source or spring of order originally within itself as well as mind does; and there is no more difficulty in conceiving, that the several elements, from an internal unknown cause, may fall into the most exquisite arrangement, than to conceive that their ideas, in the great universal mind, from a like internal unknown cause, fall into that arrangement. The equal possibility of both these suppositions is allowed. But, by experience, we find, (according to Cleanthes) that there is a difference between them. Throw several pieces of steel together, without shape or form; they will never arrange themselves so as to compose a watch. Stone, and mortar, and wood, without an architect, never erect a house. But the ideas in a human mind, we see, by an unknown, inexplicable economy, arrange themselves so as to form the plan of a watch or house. Experience, therefore, proves, that there is an original principle of order in mind, not in matter. From similar effects we infer similar causes. The adjustment of means to ends is alike in the universe, as in a machine of human contrivance. The causes, therefore, must be resembling. (Pike 1970, 25–26)

This passage is worthy of scrutiny for what follows, because we *do* see complexity, order, and purpose in nature. And there is indeed a hard-to-shake intuition that these phenomena could not possibly arise from matter guided only by unintelligent natural causes.

We will see in the next chapter that Darwin discovered a natural causal mechanism (one unknown to Hume) that was indeed capable of explaining some of the order, complexity, and adaptation that we see in the world, thereby offering an explanation in terms of unintelligent natural causes for that which had hitherto seemed to require an explanation in terms of the operation of a supernatural intelligence. In that chapter we will see that natural selection, the mechanism that brings about the emergence of functional structures and processes known as *adaptations*, is a mechanism capable of explaining, without the operation or intervention of intelligence, some of the very structures that seem to call out for intelligent design.

In point of fact, the natural, evolutionary processes giving rise to adaptations are so well documented today that many creationists will tell you that they accept *microevolution* (adaptive evolution within a species) but that they do not accept *macroevolution* (evolution giving rise to new species). However, by accepting the scientific explanation for microevolution, modern creationists, ignorant of the history of their own arguments, concede to the evolutionists the correctness of evolutionary explanations of adaptive, functional structures and processes by natural, unintelligent causes. Yet it was these very same functional structures and processes that were supposed to establish the need for intelligent causation as a consequence of the argument from design. It is no accident, in the light of Darwin's success, that a contemporary intelligent design theorist like Michael Behe has searched long and hard to try to find adaptive, functional structures and processes (alleged to be lurking in the biochemistry of organisms) that seem to resist a Darwinian explanation. His arguments will be examined in later chapters.

There is another point here. Natural evolutionary causes, important as they are, cannot account for all the order and complexity we see in nature. Natural selection does not operate on inanimate objects. Though astronomers talk of stellar evolution, they do not mean that the stars literally evolve, as do populations of organisms. *Evolution* is a word with many meanings, and we must be careful not to confuse them. Nevertheless, inanimate objects do organize themselves in certain circumstances, and without the intervention of a designing intelligence, into complex, ordered structures. For example, natural gravitational mechanisms, operating in accord with the laws of physics, can account for the ways in which stars in galaxies become organized into enormous spiral structures.

Other natural causal mechanisms can account for complexity and organization as it is observed in complex systems (ranging from the molecular to the stellar) in the world around us. Scientists discuss these causal mechanisms, operating in accord with the laws of nature, under the heading of self-organization and self-assembly. Some of these phenomena are of great interest to polymer scientists, biologists, materials scientists, and engineers. *Self-organization* is a phenomenon involving the coordinated action of independent entities (molecules, cells, organisms, or stars, perhaps) lacking centralized control (intelligent or otherwise) but operating and interacting with each other in accord with natural mechanisms to produce larger structures or to achieve some effects reflecting group action. We will discuss self-organization in later chapters. It suffices for the present purposes to note that nature, be it at the level of molecules, organisms, or stars, has natural organizing power arising from its very constitution as matter and energy.

Thus, if you throw several pieces of wood together, you won't see the pieces self-organize into a house. The conditions are not right, and you would do better here to hire an intelligent architect and a reasonably smart (and sober) group of builders. But if the architect is stupid and the builders are drunk, once again, you won't get a house. The conditions are not right. By contrast, protein molecules can self-organize into structures like the microtubules that are found in your cells; individual cells in a developing animal interact with other cells, differentiate as a consequence, and self-organize into the tissues that will give rise to its organs. Individual organisms such as insects who are members of certain species of termites, wasps, and ants, though lacking intelligence, interact with each other physically and chemically in such a way as to self-organize into a collective whose group behaviors can fashion elaborate termite mounds, wasps' nests, or ant colonies. And stars, also lacking in intelligence, interact through exchanges of gravitational energy and in the process self-organize into the mighty spiral structures observable to astronomers, all without deliberate, intentional, intelligent guidance.

Hume was unacquainted with the mechanisms giving rise to the organizing power of matter. But he was acquainted with someone who had early insights into the ways in which systems with many interacting parts can best organize without intelligent guidance into something beneficial to the group as a whole—something with valuable, functional properties that was capable of adapting to changing circumstances. The acquaintance was the great Scottish economist, Adam Smith, who was a professor at the University of Glasgow and whose *Wealth of Nations,* published in 1776, is the classical cornerstone of capitalist free market economics.

For Adam Smith (and many smart folk since), markets do best if they are left to their own devices, without centralized intelligent design and manipulation by government. As Adam Smith observed:

> It is not from the benevolence of the butcher, the brewer or the baker that we expect our daily bread, but from their regard to their own self interest. . . . [Every individual] intends only his own security, only his own gain. *And he is in this led by an invisible hand* to promote an end that was no part of his intention. By pursuing his own self interest, he frequently promotes that of society more effectually than when he intends to promote it. (quoted in Dixit and Nalebuff 1991, 223, my italics)

Economies are complex systems, some of whose parts are intelligent, but whose collective action brings about good effects that no single intelligence (or, indeed, a cooperative consisting of many) deliberately designed, intended, or caused. The good effects result from self-organization—that is, the invisible hand of economic mechanisms operating in accord with the laws of supply and demand.

The hand is invisible precisely because the good effects of market mechanisms for the economy as a whole are not deliberately intended and brought about by any intelligence (or small, centralized group of such) deliberately working to that end. As biologist Thomas Seeley has recently remarked:

> The subunits in a self-organized system do not necessarily have low cognitive abilities. The subunits might possess cognitive abilities that are high in an absolute sense, but low relative to what is needed to effectively supervise a large system. A human being, for example, is an intelligent subunit in the economy of a nation, but no human possesses the information-processing abilities that are needed to be a successful central planner of a nation's economy. (2002, 316)

In chapter 3, we will meet self-organizing systems whose subunits are cognitively vacant molecules but that nevertheless work together to produce highly ordered and organized states of matter.

The lesson here is this: Something as functional and adaptive as a market economy that looks as if it must be the result of centralized intelligent design and control is in reality nothing of the sort. Appearing as if it is intelligently designed to bring about the common good does not imply that it literally is so designed. Indeed, our experience with centralized intelligent design and control of economic systems, such as those found in numerous disastrous experiments with socialism in the twentieth century, contains parables worth heeding by the erstwhile champions of intelligent design in nature.

But this brings me back to Hume. For not only has intelligent design been disastrous in the context of economics but also there is much in nature that does not seem to be designed well at all. No intelligent, sensible, and benevolent engineer would have designed humans to be so subject to diseases like cancer; such a benevolent engineer would surely not have designed pathogens so adapted to our bodies and effective at making us sick. Surely only a buffoon or a malicious intelligence would have designed the human lower back to be the source of so much pain, and no sensible engineer would have come up with a system for childbirth as difficult and painful as that found in humans.

In this light we can perhaps appreciate the words of Hume in his own discussion of the puzzles raised by natural theology:

> In a word, Cleanthes, a man who follows your hypothesis is able perhaps to assert, or conjecture, that the universe, sometime, arose from something like design: but beyond that position he cannot ascertain one single circumstance; and is left afterwards to fix every point of his theology by the utmost license of fancy and hypothesis. This world, for aught he knows, is very faulty and imperfect, compared to a superior standard; and was only the first rude essay of some infant deity, who afterwards abandoned it, ashamed of his lame performance: it is the work only of some dependent, inferior deity; and is the object of derision to his superiors: it is the production of old age and dotage in some superannuated deity; and ever since his death, has run on at adventures, from the first impulse and active force which it received from him. (Pike 1970, 55)

If we set aside unwarranted speculation to the effect that this world, imperfect as it is, is the best of all possible worlds, warts and all, or if we reject the idea that the design defects are to be dismissed as mysteries beyond the scope of human ken, then we cannot count only examples that are evidences of good design and ignore all the evidence of bad design. Sauce for the goose is sauce for the gander! What do these evidences of imperfect design tell us about the hypothetical designer? Perhaps, adopting the old tactic of blaming the victims, the design defects result from original sin and are visited upon the sons of Adam and the daughters of Eve for this reason alone. Perhaps the intelligent designer was drunk, stupid, or both. We do not know, and we have no rational means of investigating, let alone settling, the matter.

However, Hume's most devastating critique of the argument from design springs from his invitation to do what all scientists have to do, and that is to consider alternatives to their own favored explanations of phenomena, if only to bolster the case for their own favored explanation through a rational rejection of the alternatives as being inferior in various relevant respects. In the case of the argument from design, we have the analogy of complex adaptive structures arising from the intelligent design of a craftsman. But no human craftsman has ever made an organism, much less a universe. Animals make other animals, however. So why not consider animal reproduction as an analogy for the way the universe came into being? No animal has made a universe either, but animals do make other animals, including complex intelligent animals such as ourselves.

So can we make a parallel to the argument from design that we might term the argument from animal reproduction? Hume evidently thought so:

> Compare, I beseech you, the consequences on both sides. The world, say I, resembles an animal; therefore it is an animal, therefore it arose from generation. The steps, I confess, are wide; yet there is some small appearance of analogy in each step. The world, says Cleanthes, resembles a machine; therefore it is a machine, therefore it arose from design. The steps are here equally wide, and the analogy less striking. And if he pretends to carry on my hypothesis a step further, and to infer design or reason from the great principle of generation, on which I insist; I may, with better authority, use the same freedom to push further his hypothesis, and infer a divine generation or theogony from his principle of reason. I have at least some faint shadow of experience, which is the utmost that can ever be attained in the present subject. Reason, in innumerable instances, is observed to arise from the principle of generation, and never to arise from any other principle. (Pike 1970, 65)

And just as the argument from design has an ancient ancestry in heathenism, so, too, does the argument from animal reproduction:

> Hesiod, and all the ancient mythologists, were so struck with this analogy, that they universally explained the origin of nature from an animal birth, and copulation. Plato too, so far as he is intelligible, seems to have adopted some such notion in his Timaeus. The Brahmins assert, that the world arose from an infinite spider, who spun this whole complicated mass from his bowels, and annihilates afterwards the whole or any part of it, by absorbing it again, and resolving it into his own essence. Here is a species of cosmogony, which appears to us ridiculous; because a spider is a little contemptible animal, whose operations we are never likely to take for a model of the whole universe. But still here is a new species of analogy, even in our globe. And

> were there a planet wholly inhabited by spiders, (which is very possible), this inference would there appear as natural and irrefragable as that which in our planet ascribes the origin of all things to design and intelligence, as explained by Cleanthes. Why an orderly system may not be spun from the belly as well as from the brain, it will be difficult for him to give a satisfactory reason. (Pike 1970, 66–67)

Once again, evidential sauce for the goose is sauce for the gander. The intelligent design theorist says that animals, including our intelligent selves, are the fruits of intelligent design on a cosmic scale. The animal reproduction theorist says that intelligence is only found in animals like ourselves (and by analogy other things) that result from prior animal reproduction. Chickens and eggs!

One theorist jumps to the conclusion that the entire universe, including animals, results from the designing machinations of a cosmic intelligence. The other jumps with equal alacrity to the conclusion that the universe, including animals and such other intelligences as there are, results from the reproductive operations of a cosmic creature pregnant with worlds. (An imaginative person could no doubt come up with many other alternatives, neither better nor worse than that of the design theorist.) The point is not to settle this issue one way or the other but to ask how, in the nature of the case, it could be settled. What experiments, what evidence would we need? How would we proceed to deal with this issue? These are the sorts of questions that are prompted by Hume's analysis of the argument from design.

I have dragged you through this lengthy discussion of the argument from design to show you that it has played a long role in debates about the nature of the world we live in. In Hume and Kant's day, one could be respectful of the argument from design and critical at the same time. When I look at the argument from design in the social, religious, and scientific context in which it breathed and lived and animated discussions of nature, I, too, am respectful. In fact, the natural theologians of the seventeenth and eighteenth centuries were daring, magnificent, and sophisticated thinkers. They were giants adapted as well to the intellectual and scientific environment in which they were embedded as the equally magnificent dinosaurs were superbly adapted to the environment of the ancient Cretaceous, more than 65 million years ago. But I am not respectful of the argument as it appears today in the hands of modern creationists who lack the intellectual rigor and curiosity of the natural theologians of old from whom they descend and who wish to turn back the clock of science to earlier, ignorant times.

Thus, to pursue the comparison with dinosaurs still further, the dinosaurs, wonderful as they were, became extinct owing to a meteor impact that radically altered the conditions of life, leaving most of them without a place in the economy of nature—while the survivors that did find a place were already on the evolutionary path that would lead to modern birds. Similarly, the meteoric impact of Darwinism radically altered the conditions of science. The consequent changes in our understanding of organic nature would be as telling for natural theology as the changes wrought by a more literal meteor were for the dinosaurs. The natural theologians surviving this impact evolved into creation scientists, who, like the birds the dinosaurs became, also have a place in the contemporary economy of knowledge, but only, alas, as parasites crawling on the body of science. And so at last I turn to Darwin.

2

Darwin and the Illusion of Intelligent Design

We have now traced the roots of the argument from design. There are two versions of the argument. One calls for intelligent design of the entire universe, whereas the other justifies the appeal to intelligent design by pointing to adaptive functional structures and processes observed in organisms. These arguments will be considered separately. Charles Darwin (1809–1882) responded to the argument from design that proceeds from the appeal to adaptive, functional structures in organisms. We will see that he argued that these structures and processes can be accounted for in terms of natural, unintelligent, unguided mechanisms—mechanisms that scientists could study.

Darwin's theory of evolution was but one of a series of evolutionary theories that had been proposed in the eighteenth and nineteenth centuries. Darwin's theory is important because it contains an explicit statement of how a natural, unguided mechanism, operating in accord with the laws of nature, could bring about the structures and processes that others, such as Paley, believed could be explained only as a result of intelligent, supernatural causes. But evolution has evolved considerably since Darwin's day. Accordingly, it will also be useful in this chapter to give some consideration to these more recent developments, which will help us understand issues discussed in later chapters.

Darwin and the Rock of Ages

Given allegations by religious extremists that Darwin was in league with the Devil, perhaps that he was an enemy of God, or that he was merely an atheist, the question naturally arises as to his views on religion. The issue is not quite as simple as it might seem. For example, Darwin ended the sixth edition of *The Origin of Species* (first published in 1859) with the following remarks: "There is a grandeur in this view of life, with its several powers, having been originally breathed by the Creator into a few forms or into one; and that, whilst this planet has gone cycling on according to the fixed law of gravity, from so simple a beginning endless forms most beautiful and most wonderful have been, and are being evolved" (1970, 123). Perhaps, then, he believed, like some deists, that God created life (and the rest of the world) but then left it alone to run in accord with natural processes. Perhaps, like theistic evolutionists, Darwin was suggesting that God initiated life and has been doing his work ever since by guiding the process of evolution—a process that appears to be mindless but is in reality guided by an invisible supernatural hand, we know not how or why.

Then again, perhaps Darwin, not knowing much chemistry (biochemistry at this time was an unborn fetus in the minds of scientists), felt that God could go in this gap in human knowledge. Or perhaps he did not really care about the issue of the origin of life. What matters to the evolutionary biologist is what happens after life has been initiated—ours is not to reason how or why it all came about; ours is only to explain the changes we see around us. God then serves as a convenient metaphor to explain an origin beyond the purview of the evolutionary biologist. Perhaps, as some scholars have suggested, he left in the reference to God as a sop to his wife, who was a committed Christian.

To help shed some light on these matters, it will be useful to consider other remarks Darwin made on the topic of religion, and to these we now turn. As we will see, even here there is an interesting evolutionary story to be told. In this process, we will learn that truth is indeed stranger than fiction. Darwin was a true believer when he sailed away from England on his voyage of discovery on HMS *Beagle*, even generating amusement among the ship's officers for quoting the Bible to settle moral debates. After the voyage ended, things began to change. In his *Autobiography*, he wrote:

> By further reflecting that the clearest evidence would be requisite to make any sane man believe in the miracles by which Christianity is supported,—and that the more we know of the fixed laws of nature the more incredible do miracles become,—that the men at that time were ignorant and credulous to a degree almost incomprehensible to us,—that the Gospels cannot be proved to have been written simultaneously with the events,—that they differ in many important details, far too important, as it seemed to me, to be admitted as the usual inaccuracies of eyewitnesses;—by such reflections as these . . . I gradually came to disbelieve in Christianity *as a divine revelation*. The fact that many false religions have spread over large portions of the earth like wild-fire had some weight with me. (F. Darwin 1888, 1:278, my italics)

But as Darwin was aware, one might question the literal truth of the Bible and nevertheless accept the argument from design.

After all, there are Christians today who do not see the Bible as being literally true but who are impressed by the design argument. Some proponents of intelligent design theory fall into this category. Moreover, we have seen that heathens were impressed by the argument from design before Christianity appeared on the face of the earth. (It goes without saying that there are also Christians who are content with the findings of evolutionary biologists and who reject the argument from design, lock, stock, and barrel. Fundamentalists call these latter folk *liberals*.) So what factors inclined Darwin to skepticism concerning the religion of his raising and its reliance on the biological argument from design?

Darwin was evidently swayed by versions of the argument from evil as an argument against the existence of God. In a nutshell, the argument questions whether it makes sense to suppose that an all-powerful, all-knowing, everywhere present, and completely good God would allow suffering—that is, evil—to exist in the created world. As Darwin observed:

> That there is much suffering in the world no one disputes. Some have attempted to explain this with reference to man by imagining that it serves for his moral improvement. But the number of men in the world is as nothing compared with that of all other sentient beings, and they often suffer greatly without any moral improvement. This very old argument from the existence of suffering against the existence of an intelligent First Cause seems to me a strong one; whereas . . . the presence of much suffering agrees well with the view that all organic

> beings have been developed through variation and natural selection. (F. Darwin 1888, 1:280–281)

Darwin does not here praise suffering but merely points out that its existence is a factual part of the organic predicament, and it is, moreover, what might be expected from the operation of the unintelligent mechanism (neither good nor bad but merely indifferent) that he had proposed to explain the changes we see around us.

Where others saw intelligent, beneficent design, Darwin saw misery, and it weighed upon him, and he evidently struggled with it, as can be seen in the following remarks to his friend Asa Gray in 1860:

> I cannot persuade myself that a beneficent God would have designedly created the Ichneumonidae [parasitic insects] with the express intention of their feeding within the living bodies of Caterpillars, or that a cat should play with mice. Not believing this, I see no necessity in the belief that the eye was expressly designed. *On the other hand, I cannot anyhow be contented to view this wonderful universe, and especially the nature of man, and to conclude that everything is the result of brute force.* (F. Darwin 1888, 2:105, my italics)

Darwin, working within the theological framework of his early religious training, did not know what to do with this conflict. He even suggested that the matter may be beyond human ken: "A dog might as well speculate on the mind of Newton. Let each man hope and believe what he can." But he clearly thought that his views in the *Origin of Species* were not necessarily an expression of atheism, for as he added in his letter to Gray:

> The lightning kills a man, whether a good one or a bad one, owing to the excessively complex action of natural laws. A child (who may turn out to be an idiot) is born by the action of even more complex laws, and I can see no reason why a man, or other animal, may not have been aboriginally produced by other laws, and that all these laws may have been expressly designed by an omniscient Creator, who foresaw every future event and consequence. But the more I think, the more bewildered I become. (F. Darwin 1888, 2:105–106)

While I suspect that Darwin was an atheist when he died, and that the Christian legend of a deathbed conversion bears false witness against a dead man who could no longer defend himself, it is clear that he was a complex and subtle thinker, not the simple, matter-thumping materialist of caricature. In 1860, Darwin's skepticism was not the kind that involved active disbelief; it was very much in the ancient spirit of a skepticism that involved the withholding of final judgment one way or the other.

Yet even this passage can be reinterpreted. To some, it has signaled the covert atheism of a man attempting damage control after the publication of the *Origin of Species*. To others, it has signaled that he saw the need for a more sophisticated theodicy (account of God's relation to the world) than was suggested in biological versions of the argument from design. On both reinterpretations, Darwin rejects the argument from design, but in the former he rejects it lock, stock, and barrel, both from a biological and a cosmological point of view. In the latter view, he merely rejects design in the form of an invisible supernatural hand guiding biological events, while either accepting, or at least leaving it open, as to whether there was a more primal cosmological design. Recently, Kenneth Miller (1999) has tried to develop something like this latter perspective into a rational and coherent position in his *Finding Darwin's God*, by banishing design from biology and biochemistry while retaining it for cosmology. We will meet Miller again later on in this book.

Darwin certainly tried to reach out to religious readers. Religion had adapted itself to science in the past, and perhaps it could do so here. Thus Darwin observed in *The Origin of Species*:

> I see no good reason why the views given in this volume should shock the religious feelings of any one. It is satisfactory as showing how transient such impressions are, to remember that the greatest discovery ever made by man, namely the law of attraction of gravity, was also attacked by Leibnitz, "as subversive of natural, and inferentially revealed, religion." A celebrated author and divine has written to me that "he has gradually learnt to see that it is just as noble a conception of the Deity to believe that He created a few original forms capable of self-development into other and needful forms, as to believe that He required a fresh act of creation to supply the voids caused by the action of His laws. (1970, 116)

In the last chapter, we saw that Newton had replaced the invisible hand of God with the invisible force of gravity, yet Newton was also a committed theist. That it is possible to reconcile evolution and religion in some way such as this can be acknowledged by atheists and materialists, even while they themselves are highly skeptical of such religious claims having independent warrant, or indeed any warrant at all.

But none of this should detract from the fact that when Darwin was a young man, he was a Bible-believing Christian. Thus, when Darwin set off on the famous voyage of circumnavigation on HMS *Beagle* in 1831, he was committed, as a Christian, to the correctness of the argument from design. So, when the *Beagle* had stopped in Australia, Darwin could observe in his travelogue, *Voyage of the Beagle* (1839), that the creatures in Australia, though clearly fitted into their appointed places in nature, were very different from those found elsewhere, so different as perhaps to challenge the natural theology that he had learned at Cambridge:

> An unbeliever in everything beyond his own reason might exclaim, "*Two distinct Creators must have been at work; their object, however, has been the same, and certainly the end is complete.*" While thus thinking, I observed the hollow conical pitfall of the lion-ant: first a fly fell down the treacherous slope and immediately disappeared; then came a large but unwary ant. . . . But the ant enjoyed a better fate than the fly, and escaped the fatal jaws which lay concealed at the base of the pit. There can be no doubt but that this predacious larva belongs to the same genus with the European kind, though to a different species. Now what would the sceptic say to this? *Would any two workmen ever have hit upon so beautiful, so simple, and yet so natural a contrivance? It cannot be thought so: one Hand has surely worked throughout the universe.* (C. Darwin 1839, 325, my italics)

In this way, careful study revealed to Darwin just one designing hand, where the unwary might have been tempted to see two. But Darwin would soon be led to abandon all appeals to invisible supernatural hands. In so doing, he would question the inference from organic nature's appearing to us as if it was intelligently designed, to the conclusion that it is literally intelligently designed.

The distinction between appearance and reality—between the way things are and the way they appear to be—has entertained generations of philosophy students, from those of yesteryear who worried about what sounds, if any, were made by trees falling in empty forests, to those today who are rumored to fret about unobserved refrigerator lights. The distinction between appearance and reality drawn by scientists is somewhat subtler. It is often a distinction between the way something might appear to untutored common sense and the way that phenomenon is actually generated in terms of discoverable natural mechanisms, so as to cause just that appearance.

For example, it does look as though the sun rises in the east and sinks in the west. The untutored common sense of our ancestors told them this was so and led them to conclude that the sun orbited the Earth. As we all know, there was at one time considerable resistance, not least from religious authorities in the Vatican, to a more informed view of these matters that would explain the same phenomenon as arising from the earth's rotational motions as it orbited the sun. For another example, go into a room at home and touch a piece of metal and a piece of Styrofoam (or foam rubber). The metal feels colder than the Styrofoam. But if the items have been left alone in the same environment, they are both at room temperature, and the metal feels colder because it is simply better than the Styrofoam at conducting heat away from your fingers. (For other examples, think of well-known illusions such as the appearance of a bent stick when a straight stick is partially immersed in water or the illusion of a big moon close to the horizon.) For the scientist, appearances can be misleading—but in fruitful ways—for by coming to understand how the appearances are generated, we may come to understand interesting things about the world we live in.

Lest I appear unduly harsh to untutored commonsense, science is critical of itself, too. Settled scientific wisdom—tutored common sense, if you will—about the way appearances are generated by natural mechanisms may turn out to be mistaken on the basis of a more careful analysis of natural mechanisms. Twenty years ago, it was settled medical wisdom that virtually all stomach ulcers were the result of stress, and ulcer management meant careful diet and avoidance of stress. Today we understand that many stomach ulcers result from treatable bacterial infections. Scientific advances often involve the correction and even the abandonment of earlier scientific views, even views that are held very dear by the scientists steeped in them. Darwin would ultimately come to see the evidence of design that had so impressed natural theologians such as Paley as appearances generated by the operation of natural mechanisms in accord with natural laws. But this is to look ahead.

It is only after the voyage of the *Beagle* comes to an end in 1837, and Darwin had time to reflect upon what he had observed in his capacity as a naturalist, that doubts set in. But bare observation on its own tells a scientist very little. Observations need to be interpreted and made sense of. Only then, duly interpreted and analyzed, do they become constraints upon our theorizing, supporting some of our ideas and leading us to reject others. This interpretation and analysis of observation is something that takes place against a background of theory. Our observations are inescapably contaminated by the theories we are exposed to, but the scientist is luckier than most of us, because the background theories in science have typically been tried and tested and tried again. And when found wanting, they are adjusted or rejected, often with consequent changes in how we see and interpret other things around us.

Thus to understand Darwin's work, it is important to understand that he was the beneficiary of a new method in science—one that emerges during the late eighteenth and early nineteenth centuries. This new method had been fruitful and had shaped investigations in many fields, but it has special implications for geology, its estimate of the age of the Earth, and how the present state of the Earth results from changes brought about by the operation of natural causes over long periods of time.

Geology and the Age of Rocks

The young Earth creationists of Darwin's day believed they lived on a juvenile planet that had been created by God around 4004 B.C. The current physical state of the Earth—with seashells found in rocks on mountain tops—evidently reflected the occurrence of catastrophic planetary upheavals, Noah's flood being the last. But by the end of the eighteenth century, there was a growing realization, at least in educated circles, that the Earth, though created, might be considerably older than these orthodox estimates implied. This view is known as *old Earth creationism*.

In the late eighteenth and early nineteenth centuries, a new method found its way into science, and William Whewell (1794–1866) called it the *method of gradation* (see Butts 1989). When we look around us, we see objects that are clearly very different from each other. But on closer scrutiny, one of the things we need to know is whether these seemingly very different objects belong to distinct categories (are different types or kinds of thing), or whether they are merely extreme points on a spectrum, connected by a range of intermediaries, each differing slightly from another, but so arranged as to connect the seemingly categorically distinct objects. As Whewell puts it: "To settle such questions, the *Method of Gradation* is employed; which consists in taking intermediate stages of the properties in question, so as to ascertain by experiment whether, in the transition from one class to another, we have to leap over a manifest gap, or to follow a continuous road" (Butts 1989, 240). The method was used by the physicist Michael Faraday to undermine an absolute distinction between electrical conductors and nonconductors through a consideration of semiconductors (sulfur is a poor conductor, spermaceti is better than sulfur, water is better than spermaceti, and metals better still).

But the method of gradation had important implications for the ways in which theorists who understood it examined data in the geological and biological sciences. For the present purposes, we must examine the work of the geologist Charles Lyell (1797–1875), whose *Principles of Geology* was published in three volumes between 1830 and 1833. Darwin sailed away from England with the first volume, and later volumes caught up with him on his epic voyage.

Lyell's enterprise was none other than an attempt to explain how the Earth had changed in the course of geological time by reference to causal processes that can be seen to be in operation today—for example, water erosion, wave action, freeze-thaw erosion, glaciation, deposition of sediments, wind erosion, volcanism, and earthquakes. Under the influence of Newton's work in physics, Lyell believed the laws of nature were the same everywhere in space and time. Thus, the laws operating in distant regions today, or in times long past, are the same as those that operate for us here and now.

Lyell thus believed that the principles underlying geological change in the past could not be different from those we see in operation today. The basic idea, then, is that the small, stepwise changes brought about by these causal processes can gradually accumulate over very long periods of time to result in substantial changes and hence substantial differences between things. Whewell, in a review written in 1832, called this idea *uniformitarianism*.

Where young Earth creationists saw the state of the planet today as the result of special catastrophes in the recent past, Lyell saw the geological record as the result of the action, over long periods of time, of the same kinds of natural causes that we see in operation today. Consider the clear and distinct appearance of massive differences between the objects of geological inquiry—for example, between seabeds below and mountaintops on high (sometimes with seashells embedded in them). According to Lyell's reasoning, these do not reflect absolute categorical differences but, in accord with the method of gradation, result instead from the slow accumulation of small changes occurring over very long stretches of time, brought about by natural causes of the kind amenable to study today. The differences do not result from massive catastrophes brought about by supernatural agency in very short periods of time. Given time enough, seabeds can be pushed up into mountain ranges. But time enough was not to be measured in mere thousands of years but in terms of many, many millions of years. The new scientific approach to geology was not inconsistent with old Earth creationism, but it conflicted mightily with young Earth creationism.

However, there was something else here of great importance. Lyell's uniformitarian approach to geological change carried with it the implication that the physical environment on Earth is not static. If the physical environment is slowly changing, then intelligently designed organismal machines, once fitted into their appointed places in nature, would, by staying the same over many successive generations, find themselves out of kilter with the natural environment in which they were embedded. Yet this is not what we see. Instead, we see remarkable ranges of adaptation. In his early work, Lyell believed that species became extinct as the conditions for which they were initially designed and adapted changed. The resulting gaps in nature were supposed to be refilled by the introduction of new species, presumably by supernatural means (Mayr 1991, 16). Darwin, under the influence of Lyell's writings, saw the evidence of adaptation relative to a changing environment very differently.

Darwin and the Origin of Species

Charles Darwin published the *Origin of Species* in 1859. His achievement reflects his intellectual inheritance. Darwin is an heir to the method of gradation—species differences are not absolute, categorical discontinuities. Thus two closely related species will be similar to each other in some respects and different in others—think of humans and chimpanzees. The line that leads to modern humans diverged from the line that leads to modern chimpanzees several million years ago. Evolution is thus a branching process. These two lineages diverge from the lineage of the common ancestor of human and chimpanzees. Chimpanzees thus did not evolve into humans; rather, humans and chimpanzees descended with modification from a common ancestor in the distant past that was neither human nor chimpanzee.

The similarities between humans and chimpanzees reflect this common evolutionary ancestry from a now extinct parental species. The differences between humans and chimpanzees reflect evolutionary modifications that occurred after the respective lineages diverged. Humans and chimpanzees, like Faraday's conductors and nonconductors, are connected by a series of intermediate cases. This is known as the principle of *phylogenetic continuity*. As Darwin observed in a letter to Asa Gray:

> Each new variety or species when formed will generally take the place of, and so exterminate its less well-fitted parent. This I believe to be the origin of the classification or arrangement of all organic beings at all times. These always *seem* to branch and sub-branch like a tree from a common trunk; the flourishing twigs destroying the less vigorous—the dead and lost branches rudely representing extinct genera and families. (F. Darwin 1888, 1:481)

If Darwin is right, then the taxonomic order seen in nature reflects neither intelligent design, as Paley had supposed, nor the whimsy of the taxonomist, but the historical facts resulting from the operation of evolutionary mechanisms.

In practice, modern evolutionary biologists rely on many sources of evidence to figure out evolutionary relationships. The evidence ranges from facts uncovered by comparative anatomists and comparative physiologists, to facts concerning similarities and differences at the molecular level (protein structure and gene sequences), to facts provided by the fossil record itself. I mention all this to emphasize that the fossil record is one type of evidence studied by evolutionary biologists, but it is by no means the only type of evidence. And if the fossil record is notoriously incomplete and filled with gaps (because most creatures do not fossilize, because many fossils are in the hearts of mountains awaiting discovery, or because many have been exposed and eroded away, thereby vanishing forever), fossil hunters have nevertheless made many impressive discoveries with respect to intermediate evolutionary forms (the links that connect diverging lineages). A good introduction to these matters is contained in Strahler (1989). For a startling modern vindication, through the discovery of numerous fossil intermediates, of Darwin's claim that whales descended from terrestrial mammals, see Stephen J. Gould's (1995) essay "Hooking Leviathan by Its Past."

In his later work, Darwin employed the method of gradation to address the issue of the relative extents of cognitive development in animals:

> We have seen . . . that man bears in his bodily structure clear traces of his descent from some lower form; but it may be urged that, as man differs so greatly in his mental power from all other animals, there must be some error in this conclusion. . . . If no organic being excepting man had possessed any mental power, or if his powers had been of a wholly different nature from those of the lower animals, then we should never have been able to convince ourselves that our high faculties had been gradually developed. But it can be shewn that there is no fundamental difference of this kind. We must also admit that there is a much wider interval in mental power between one of the lowest fishes, as a lamprey or lancelet, and one of the higher apes, than between an ape and man; yet this interval is filled up by numerous gradations. (1871, 65)

In other words, nature affords numerous examples of cognitive gradation. If this was right, then the seventeenth-century philosopher René Descartes was simply wrong to view all animals as cognitively vacant machines, humans differing from them by possession of a nonphysical mind or soul. Moreover, there is no reason to suppose that similar gradations did not exist in the lineages leading to modern chimpanzees and modern humans, respectively, after their divergence from a common ancestor (now estimated to be more than seven million years ago).

Today, with a better understanding of the fossil record than Darwin had, we have found evidence of such cognitive gradation. This can be seen in the evidence of increasing cranial capacity in the various species found along the line that leads to modern humans. In the last three and a half million years, there has been a substantial increase in average cranial capacity, from about 550 cubic centimeters in our australopithecine ancestors to about 1,200 cubic centimeters in modern humans. And there is further cultural evidence in terms of growing sophistication in toolmaking and tool-using skills (see Park 1996; Shanks 2002).

But Darwin's rationale for the search for evidence of cognitive similarity, regardless of whether his own methods were up to the task at hand, reflects the consequences of taking evolution seriously. In 1872, Darwin published *The Expression of the Emotions in Man and Animals*. In this book, he comments on the scientific sterility of viewing emotional expression in humans and other animals in terms of the argument from design: "No doubt as long as man and all other animals are viewed as independent creations, an effectual stop is put to our natural desire to investigate as far as possible the causes of Expression. By this doctrine, anything and everything can be equally well explained; and it has proved as pernicious with respect to Expression as to every other branch of natural history" (1965, 12). By contrast, adherence to the method of gradation inclines an investigator to conduct comparative studies, looking for differences and similarities with respect to cognition and emotional expression between members of distinct species.

The investigator will examine cognitive adaptations and try to locate them in ecological context (how they enable an organism to make its living) and in evolutionary context (how these adaptations contribute to reproductive success). As Darwin put it:

> With mankind some expressions, such as the bristling of the hair under the influence of extreme terror . . . can hardly be understood, except on the belief that man once existed in a much lower and animal-like condition. The community of certain expressions in distinct though allied species, as in the movements of the same facial muscles during laughter by man and by various monkeys, is rendered somewhat more intelligible, if we believe in their descent from a common progenitor. (1965, 12)

These studies make sense on the assumption that the subjects of these comparative inquiries bear evolutionary relationships through descent from common ancestors with subsequent evolutionary modification. On this view, organisms carry the legacies of their evolutionary histories and relationships with them.

If, by contrast, the subjects of these inquiries are designed independently, we know not how, so that we cannot determine whether similarities and differences are intentional on the part of the designer or merely accidental, then there is no expectation that fruitful discoveries might be made about one of them by studying another. Unlike the human design process, where we can study craftsmen and test hypotheses about their methods and intentions (e.g., if the craftsman makes musical instruments, were similar techniques of design and construction used in making cellos and guitars as were used in the manufacture of violins?), supernatural design is utterly opaque and beyond the hope of rational inquiry. Issues related to this point will resurface later in the context of contemporary intelligent design arguments.

But Darwin has more than the method of gradation. He is also a beneficiary of Lyell's uniformitarianism. Thus, the differences we see between species today reflect the slow accumulation, over long periods of time, of small changes brought about by unguided, natural causes similar to those we see in operation today or, as Darwin himself puts it:

> But the chief cause of our natural unwillingness to admit that one species has given birth to clear and distinct species, is that we are always slow in admitting great changes of which we do not see the steps. The difficulty is the same as that felt by many geologists, when Lyell first insisted that long lines of inland cliffs had been formed, the great valleys excavated, *by the agencies which we still see at work*. The mind cannot possibly grasp the full meaning of the term of even a million years; it cannot add up and perceive the full effects of many slight variations accumulated during an almost infinite number of generations. (1970, 116–117, my italics)

Evidence suggested that plants and animals had adapted to environmental changes, but prior to Darwin, there was no really good explanation for how these changes occurred.

Darwin's crucial insight was to consider the problem from the standpoint of populations. First of all, individuals come and go, but populations typically exist for many generations. Individuals live and die, reproducing if they are lucky, but they do not evolve. Populations of individuals evolve over time. Evolution thus occurs across generations, and its pace is governed in part by generation time, which in humans is about twenty years but in a microorganism like *Staphylococcus aureus* may be as little as twenty minutes. One effect of evolution is to gradually change the way in which a population of organisms is structured—in particular, with respect to the statistical frequencies of characteristics that are found in individuals making up the population. But what mechanism could bring about such effects in populations over many successive generations?

Darwin observed that members of natural populations of organisms typically show variation with respect to heritable traits. Since the dawn of agriculture, animal breeders had long exploited naturally occurring intraspecific (within species) variation to make new varieties: Only animals with desirable traits (wooliness of coat, milk yield, domesticity, etc.) were allowed to reproduce and pass these traits on to the next generation, where the process would be repeated. Over time, animal breeders were able to change the way in which domestic populations of animals were structured. But natural varieties—Darwin called them *incipient species* (1970, 39)—pervade nature and not just the farmer's yard.

Without appealing to supernatural intelligent design, how could nature work with the variation in heritable traits found in natural populations to bring about adaptations and ultimately the origin of new species? What natural mechanisms might be at work to this end? Darwin's answer reflects his acquaintance with some ideas originally explained by Thomas Malthus (1766–1834) in his *First Essay on Population* (1798). According to Malthus, much human misery arises from the tendency of populations to grow faster than they can increase food supply to support their numbers. Starvation, conflict, and disease are the consequences of this process, and they are consequences whose effects trim expanding populations back, changing their structure in the process.

As applied to natural populations generally, this suggested to Darwin that a struggle for existence arises naturally from the fact that organisms tend to produce more offspring than can be supported by the environment: "Every being, which during its natural lifetime produces several eggs or seeds, must suffer destruction during some period of its life, and during some season or occasional year, otherwise, on the principle of geometrical increase, its numbers would quickly become so inordinately great that no country could support the product" (1970, 41). It is here, in the context of the superabundance of organisms, that heritable variation plays its crucial role.

In this ongoing struggle for existence, some organisms—variants—will have characteristics that hamper their ability to survive and reproduce; other variants will have characteristics that enhance these same abilities. Such traits aiding survival and reproduction are said to confer fitness advantages: "Owing to this struggle, variations, however slight and from whatever cause proceeding, if they be in any degree profitable to the individuals of a species, in their infinitely complex relations to other organic beings and to the physical conditions of life, will tend to the preservation of such individuals, and will generally be inherited by the offspring" (1970, 39). For Darwin, this mechanism is the primary engine of evolution:

> This preservation of favorable individual differences and variations, and the destruction of those which are injurious, I have called *Natural Selection*, or the *survival of the fittest*. Variations neither useful nor injurious would not be affected by natural selection, and would be left either a fluctuating element, as perhaps we see in certain polymorphic species, or would ultimately become fixed, owing to the nature of the organism and the nature of the conditions. (1970, 44)

Natural selection thus works on heritable variation found in populations of organisms. In the environment in which the struggle for existence takes place, the traits favored by selection increase in frequency over successive generations, and they come to represent adaptations to the environment in which the struggle for existence occurs. Adaptations are those features of organisms that are the quintessential fruits of the operation of natural selection.

Importantly for our purposes, it was Darwin's contention that the same selective mechanisms that bring about adaptations *within* populations of organisms will also, as an unintended by-product, gradually bring about and amplify differences *between* populations great enough to constitute their designation as separate species. In this way, Darwin's understanding of species differences reflects the method of gradation. What appear to be absolute categorical differences turn out instead to be extreme differences that arose gradually by degrees through the accentuation of differences between varieties of a given species. As Darwin put it:

> On the view that species are only strongly marked and permanent varieties, and that each species first existed as a variety, we can see why it is that no line of demarcation can be drawn between species, commonly supposed to have been produced by special acts of creation, and varieties which are acknowledged to have been produced by secondary laws. On this same view we can understand how it is that in a region where many species of a genus have been produced, and

> where they now flourish, these same species should present many varieties; for where the manufactory of species has been active, we might expect, as a general rule, to find it still in action; and this is the case if varieties are incipient species. (1970, 108)

In this way, the varieties that result from microevolutionary processes (processes driving changes within species that creationists have been forced to accept on pain of looking as silly as flat-Earth geographers) are driven still further apart by the continued action of the same mechanisms so as to constitute new species in their own right. In this way, microevolutionary changes, continued long enough, give rise to macroevolutionary phenomena.

An analogy might be helpful here. Sir Isaac Newton knew that objects such as cannonballs (and other bodies falling near the surface of the Earth) described parabolic trajectories (to see such a trajectory, throw a baseball to a friend on a windless day). He also knew that objects such as planets described elliptical orbital trajectories around the sun. Ellipses are very different in shape from parabolas—so different, in fact, that scientists prior to Newton believed that objects near the surface of the Earth obeyed one set of laws, while those in the heavens operated by different principles altogether. Both the trajectories and the laws describing them were viewed as being categorically different.

To undermine this view, Newton imagined there was a cannon on a mountaintop that was firing cannonballs with successively larger charges of gunpowder. The cannonballs describe parabolic trajectories in which the successive balls travel farther and farther downrange. Eventually, the gradual continuation of this process results in a truly long shot, and, while the cannonball falls to Earth, it is traveling so far and so fast that the curved surface of the Earth falls away from the cannonball. The cannonball has gone into orbit around the Earth. Newton realized that the same principles governing the way objects fall close to the Earth also apply to objects in the heavens, notwithstanding the marked differences in the behavior of those objects.

Darwin's view of the origin of new species is similar in kind. The processes driving the origin and accentuation of varieties within a species, given enough time, will turn varieties into good and true species in their own right. Over long periods, these processes result in increasing biodiversity:

> As each species tends by its geometrical rate of reproduction to increase inordinately in number; and as the modified descendants of each species will be enabled to increase by as much as they become more diversified in habits and structure, so as to be able to seize on many and widely different places in the economy of nature, there will be a constant tendency of natural selection to preserve the most divergent offspring of any one species. Hence, during a long continued course of modification, the slight differences characteristic of varieties of the same species, tend to be augmented into the greater differences characteristic of species of the same genus. (1970, 108)

This process of adaptive radiation explains what happens when animals from an ancestral species move into a multiplicity of ecological niches, each niche being characterized by a particular complex of features that affect an animal's way of making a living: nature and availability of food, type and number of predators, pathogens and parasites, climates, and so on. Thus, as small differences between populations accumulate through adaptive specialization over many successive generations, the invisible hand of natural selection will accentuate differences between these populations until they are so distinct as to be recognized as different species.

In short, to use Henry Petroski's useful turn of phrase, "Form follows failure" (1994, 22). Many are tried in the court of natural selection that a few may succeed. Evolution by natural selection is an unintelligent wasteful process, but it gets the job done, for it is a natural process whereby populations of organisms can change their characteristics over time and thus remain adapted and functional in an environment that is changing with them. The biological world in which this wasteful process takes place is the very antithesis of a well-oiled, well-designed machine with organic wheels within wheels, all turning together harmoniously that each has a natural place in the economy of nature. If Darwin is right, the economy of nature is perhaps better thought of as being analogous to a free market economy.

In the long course of the twentieth century, biologists and economists have learned much from each other. Biologists, struggling to understand the ways in which populations of organisms have evolved over time, have found many useful economic metaphors to aid them in this endeavor. Today, for example, one can find evolutionary geneticists talking about cost-benefit analyses associated with reproductive strategies (where the costs and benefits are measured in terms of offspring produced). Ideas about division of labor in the economy of nature have also resonated in the minds of biologists as they struggled to comprehend ecological specialization.

While it is hardly surprising that metaphors derived from considerations of free markets, where many individuals compete with each other for profit, can indeed be used to shed light on Darwin's ideas, we also know from the last chapter that the uncritical use of metaphors can be dangerous and misleading. Accordingly, the metaphors drawn from experience with the marketplace are merely aids to understanding. I do not use them to praise or condemn free market economics but to shed explanatory light on evolutionary processes.

In the last chapter, in a discussion of self-organization, we met the butcher, the baker, and the brewer in Adam Smith's free market economy. These individuals pursued their own economic self-interest with no larger view to the public good. In so doing, they brought about unintended beneficial effects for society at large. Competition, for example, forces competitors to be more efficient in the production of goods and can thus drive down prices, which is beneficial to consumers. These effects happen as a consequence of economic mechanisms operating in accord with the laws of supply and demand. The beneficial effects for society at large are not the result of deliberate intelligent design by any individual or group of such, natural or supernatural. The beneficial effects arise purely as unintended consequences of behaviors by individuals directed to other ends and purposes—that is, the selfish maximization of profits.

The adaptive properties of a free market appear as if they result from the hand of a designing intelligence. But there is literally no such intelligence. The unseen hand behind the appearances is found in the blind, uncaring, unintelligent market mechanisms that simultaneously govern and reflect the behavior of individual competitors. Free market economies achieve these beneficial effects in ways that are wasteful with respect to individual competitors. Small differences between competitors translate into differences with respect to profitability. Unprofitable competitors go out of business, those with a competitive edge proliferate, and in the process, markets as a whole change their structure and adapt to changing economic circumstances.

By a similar light, individual organisms in a population, each differing slightly from another in ways that are heritable, pursue their own reproductive interests. Through the mechanism of natural selection, operating in accord with the laws of inheritance, these individuals inadvertently bring about changes in the way populations are structured, with some populations flourishing at the expense of others. In a sense, evolution is pure demographics. It is about different individuals leaving behind different numbers of offspring by virtue of the possession of characteristics that can be inherited by their offspring.

For example, we humans are currently in the midst of a healthcare crisis caused by the spread of antibiotic resistance in bacterial populations. Natural bacterial populations contain heritable variation with respect to individual susceptibility to antibiotics. Most susceptible bacteria are wiped out by antibiotics and do not get to reproduce. But among the bacteria surviving the therapeutic assault with antibiotics are those who by luck of their constitution can tolerate clinical doses of a given antibiotic. These are the bacteria that survive and go on to reproduce, with their offspring inheriting a constitution tolerant of clinical doses of antibiotics. Bacterial populations thus change their structure in this way, over many generations, so as to contain many individuals who are resistant to antibiotics. In this way, drugs we had intelligently designed are rendered obsolete by evolution through natural selection.

If a natural population flourishes, this effect results from the invisible hand of the operation of blind, uncaring, unintelligent natural mechanisms in accord with the laws of nature. An individual in a natural population may appear as if it is intelligently designed, but for Darwin, appearances can be deceiving. As Darwin observes in *The Origin of Species*:

> We behold the face of nature bright with gladness, we often see the superabundance of food; we do not see or we forget, that the birds which are idly singing around us mostly live on insects or seeds, and are thus constantly destroying life; or we forget how largely these songsters, or their eggs, or their nestlings, are destroyed by birds and beasts of prey; we do not always bear in mind, that, though food may be now superabundant, it is not so at all seasons of each recurring year. (1970, 40)

Unlike the invisible hand of the supernatural intelligent designer, the invisible hand of natural selection can be seen, studied, and understood if we look hard enough at nature. It is invisible only to those who are incapable of getting behind the superficial appearances to the observable mechanisms that generate those appearances.

We saw in the last chapter that Fracastorius thought the motions of the heart were a mystery known only unto God. By careful observation of hearts from a variety of species, William Harvey unraveled that mystery. He came to see that which was invisible to Fracastorius. Darwin's achievement is similar to that of Harvey. Darwin saw what Paley had missed. As Darwin would later observe in his *Autobiography*:

> The old argument of design in nature as given by Paley, which formerly seemed to me so conclusive, fails, now that the law of natural selection has been discovered. We can no longer argue that, for instance, the beautiful hinge of a bivalve shell must have been made by an intelligent being, like the hinge of a door by man. There seems to be no more design in the variability of organic beings and in the action of natural selection, than in the course which the wind blows. (F. Darwin 1888, 1:279)

The wind is a natural phenomenon arising from natural causes. Yet even from such a humble phenomenon springs creative power. For who among us has not seen pictures of the rolling dunes of the desert that result from wind action, or the weird and wonderful mesas sculpted by wind erosion? Even the wind itself can be organized, as is shown by the mighty spiral structures of hurricanes, hundreds of miles across, as seen from outer space.

And in this observation is a glimpse of the significance of the Darwinian revolution: Evolution is a causal process but not one that fits and coheres with a view of the universe as an intelligently designed machine. The functional, adaptive properties of organisms result from what medieval philosophers would have called efficient causes. There are no final causes and hence no march of progress directed to future ends. Darwin's theory thus represents a challenge not merely to a long theological tradition but also to a way of thinking about the objects of biological inquiry—that is, organisms as mechanical components of nature's grand machine.

So much for Darwin. But evolutionary biology has itself undergone much evolution in the time since the death of Darwin. I will finish this chapter with an examination of some of these developments.

Evolution after Darwin

Modern evolutionary biologists do not believe or have faith in the literal, inerrant truth of Darwin's works any more than modern astronomers believe or have faith in the literal, inerrant truth of the works of Copernicus. Scientists and other reasonable folk can recognize the importance and significance of scientific ideas, especially as they occur in historical context, without subscribing to them as literal truths or articles of faith, let alone as revelations from the Devil. The modern biologist sees Darwin as having taken important first steps toward an evidentially grounded scientific explanation of the structures, processes, and changes we see in the biological world around us.

An example may help. Copernicus, in putting the sun at the center of the solar system, with planets describing circular orbits around it, took similar steps. But there were things such as elliptical planetary orbits of which Copernicus was unaware. Tycho Brahe gathered the data, but the explanation and interpretation of the data was left to Kepler. Kepler understood elliptical planetary orbits, and Galileo understood parabolic trajectories taken by terrestrial objects, such as cannonballs, that fall near the earth. But the principles governing celestial motion were still not fully united with those describing terrestrial motions. It would be left to Newton to unify our understandings of the motions in the heavens and the motions at or near the earth by showing that motions of both types obeyed the same dynamical laws. And in the fullness of time, it would turn out that there were things about planetary motions that Newton's laws could not adequately explain, such as the annual precession, or shift, of the perihelia of the planets (the perihelion is the point on an elliptical orbit around the sun that comes closest to the sun; the aphelion is the point farthest away). The explanation of this phenomenon would be left to Einstein and his theory of general relativity. None of this diminishes the achievements of Copernicus, for all modern astronomers are heirs to his legacy. Nevertheless, science has advanced into new explanatory territories since his day. The same is true of Darwin.

Darwin knew nothing about the mechanisms of inheritance. One of the most important developments in evolutionary biology in the twentieth century was the fusing together of evolutionary ideas about natural selection, as a force driving change in populations over successive generations, with genetics, the science of heredity and variation in populations. We will also see that, more recently still, radical new ideas about the origins of body forms are emerging from the fusing together of modern evolutionary biology with developmental biology. All these developments are helping us understand, in better and clearer ways, how organisms fit into the economy of nature—literally, how they are shaped to fit into the environments in which they are embedded.

Evolving Genes

The bringing together of Darwinian ideas about adaptive evolution by natural selection and ideas from the science of genetics concerning variation and heritability in populations results in something known as the *new synthesis* in evolutionary biology. These events took place over a thirty-year period beginning in the 1930s. Many theorists were involved in the formation of the synthesis, and the end result is a thoroughly gene-centered view of evolution. This has been popularized notably by Richard Dawkins in *The Selfish Gene* (1989).

We have just seen that for Darwin evolution was possible because of the existence of heritable variation in populations of interest. What he did not know was that cells carry genetic material. Genetics, the branch of science that deals with the nature and characteristics of genetic material, was taking its first fumbling steps while Darwin was alive. The particles of inheritance are called *genes*, and genes are made up of DNA (deoxyribonucleic acid). The distinction between an organism and its genes underlies one of the most basic distinctions in genetics, that between phenotype and genotype: "The 'phenotype' of an organism is the class of which it is a member based upon the observable physical qualities of the organism, including its morphology, physiology, and behavior at all levels of description. The 'genotype' of an organism is the class of which it is a member based upon the postulated state of its internal hereditary factors, the genes" (Lewontin 1992, 137). Corresponding to this distinction is that between genome and phenome:

> The actual physical set of inherited genes, both in the nucleus and in various cytoplasmic particles such as mitochondria and chloroplasts, make up the *genome* of an individual, and it is the description of this

> genome that determines the genotype of which the individual is a token. In like manner there is a physical *phenome*, the actual manifestation of the organism, including its morphology, physiology and behavior. (Lewontin 1992, 139)

The Genome Project has revealed that human genome contains about 30,000 genes (compared with 13,600 for the fruit fly *Drosophila*). Readers interested in learning more about genetics might consult a good textbook on undergraduate biology (e.g., Campbell 1996). Those in search of more detail would do well to consult Li (1997).

Genes are located on chromosomes, which are threadlike structures in the nucleus of a cell consisting of DNA and associated proteins. DNA consists of two chains of nucleotides, which are organic compounds consisting of a sugar (deoxyribose) linked to a nitrogen-containing base. The bases are adenine, cytosine, guanine, and thymine. The chains of nucleotides are wound around each other in the form of a spiral, ladder-shaped molecule—the famous double helix. The bases on each chain pair with a base on the other chain to form base pairs. Adenine pairs with thymine, and guanine with cytosine. Each base pair can be thought of as a bit, or basic unit, of information. There are approximately 3.5×10^9 bits of information in the human genome.

In diploid organisms—organisms with two sets of chromosomes, one from each parent (we are diploid organisms)—the matched pairs of chromosomes are called *homologous chromosomes*. In humans, barring chromosomal abnormalities, each cell contains forty-six chromosomes (twenty-two matched pairs, and one pair of sex chromosomes, with females having XX pairs and males having XY pairs). The number of pairs is called the *chromosome number*, n. In humans, $n = 23$.

The locus of a gene is its position on a chromosome. For a given locus, a population of organisms may contain two or more variant forms of the gene associated with that locus. These variant forms of a gene are called *alleles*. In diploid organisms (e.g., mammals), there are two alleles of any gene, one from each parent, which occupy the same relative position on homologous chromosomes. When one such allele is dominant and the other is said to be recessive, the dominant allele influences the particular characteristic that will appear in the organism's phenome. (It is possible for both alleles to be fully expressed; this is called *codominance*. There are also cases where neither allele is fully expressed, and a characteristic results from the partial expression of each. This is called *incomplete dominance*.)

Genes can be inherited in the form of identical copies, and these are said to be *identical by descent*. But genes passed from one generation to the next may undergo changes known as *mutations*. If a base pair changes, this is called a *point mutation*. When genes are expressed, they make proteins, which serve many functions and roles in our bodies. A point mutation in a critical location on a gene can change the nature of a protein, for good or ill, because of its implications for survival or reproduction (and hence natural selection). Because the genetic code contains redundancies, point mutations may have no effect whatsoever and so are sometimes said to be *neutral*. But when such changes occur, often more than one base pair is affected. Common changes also include the deletion of extant base pairs or the insertion of additional base pairs.

Important for our present purposes are genetic changes known as *duplications*. Entire genes can be duplicated, and when this happens the resulting genome has two copies of a gene where before it had one. Duplication events are very important for evolutionary biologists. First, because with two copies of a functional gene, one can continue its old job, while the new copy can undergo mutation and acquire new functions that can participate in the life of an organism in novel ways. This may have important implications for natural selection, by contributing to reproductive success. The process by which a gene acquires new functions in this manner is known as *exaptation*. Second, duplication is the way in which organisms acquire new genes. They do not appear by magic; they appear as the result of duplication. Duplications can also occur at the level of chromosomes and can cause serious problems. Down syndrome is a well-known result of chromosomal duplication. But entire genomes can be duplicated, with some very interesting consequences, as we will shortly see. These large-scale genomic duplications are discussed under the heading of polyploidy.

A point to bear in mind is that there is nothing good or bad in a mutation in and of itself. Instead, you must always look at the consequences of the change for the life of the organism that contains it. This will often mean an examination of the way an organism is trying to make a living in its ecological context and the challenges it faces. I'll give an example shortly. However, it is worth noting at this point that some genes are conserved. This means they have stayed the same in many lineages. What this usually means is that these genes perform essential roles in enabling the basic functions needed for life. They are the same in many lineages because mutational variants are lethal or debilitating and have been weeded out by selection. Other genes, especially duplicates, are much more tolerant of mutation and therefore can play a positive role in evolution.

Important for evolution, then, is the existence of multiple alleles in populations of organisms. A given allele may be found with a given statistical frequency in a population. Evolution occurs in a population when the relative frequency with which alleles are found in that population changes (for whatever reason) from one generation to the next. An important part of the new synthesis was the development of sophisticated techniques to analyze allele frequencies in populations. The resulting theory is thoroughly gene-centered.

By this it is meant that what gets replicated are genes, and it is genes that travel down the generations—genes, barring mutations, that are identical by descent. Underlying the heritable variation in morphological, physiological, and behavioral characteristics observed in populations is variation with respect to alleles. Parents pass on alleles to their offspring, who receive 50% of their alleles from each parent.

Recombination is the process whereby genes are shuffled during meiosis—the formation of reproductive cells (sperm or egg)—and results in offspring having a combination of characteristics different from either of their parents. By contrast, germ-line mutation is the process that results in genetic changes in an organism's reproductive cells and hence heritable changes in an organism's genetic constitution. Both these processes add to variation in populations. (There are other mutations called *somatic mutations* that result in genetic changes in cells other than the reproductive cells and that are thus not heritable. These latter mutations may have adverse effects for the organisms possessing them, such as cancer.)

What parents pass on to their offspring are alleles. Alleles that contribute positively to reproductive success are more likely to find themselves in the next generation, in higher frequencies, than alleles that do not. Such alleles are said to confer fitness advantages. Members of a population of organisms typically differ from each other with respect to their relative fitness. Differences in relative fitness are defined in terms of differential reproductive success. Thus the effect of natural selection is to change the frequencies with which alleles are found in populations over time. As Ewald has noted (1994, 4), natural selection favors characteristics of organisms that increase the passing on of the genes (alleles) that code for those particular characteristics. Evolution works across generations in populations. It is populations that evolve, not the organisms that constitute them at any given time. The phenotypic characteristics favored by natural selection are called *adaptations*.

Since we have already mentioned antibiotic resistance, consider *Staphylococcus aureus*, the microorganism responsible for much wound infection in hospitals. Such infections can be treated with antibiotics. A given population of microorganisms colonizing a patient will typically vary with respect to susceptibility to a given dose of antibiotic. *Staphylococcus aureus* reproduces asexually about every 20 minutes, giving rise to the next generation. The bacteria with alleles conferring tolerance to the clinical dose of antibiotic administered will reproduce and get those alleles into the next generation. The susceptible bacteria will be eliminated from the population. Over successive generations, alleles for antibiotic tolerance will increase in frequency. Antibiotic tolerance is thus a bacterial adaptation to hosts periodically flooded with antibiotics.

The new synthesis also resulted in an understanding of the importance of nonadaptive evolution. Allele frequencies can change for reasons unconnected with the operation of natural selection. Such changes can be effected by gene flow—the exchanges of alleles within and between populations—and the cessation of gene flow between populations can allow for the successive accumulation of significant genetic differences between those populations. Nonadaptive evolution can also result from genetic drift—changes in allele frequencies brought about by chance events. (For example, in a small population, the accidental loss of one or two individuals can bring about significant changes in the frequencies with which alleles are found; alleles that are found only in the individuals who are lost will vanish altogether from the population.) There are other ways in which nonadaptive evolution can occur, too, but the main point is that there are many ways in which allele frequencies can change, some of which involve selective mechanisms of various kinds, and some of which do not.

Evolutionary biologists believe that biodiversity results from speciation—the complex array of processes that gives rise to new species. But what exactly is a species? As we saw in the last chapter, the Aristotelian view of species held that species were groups of organisms all of which had the same *form* or *essence* despite much variation in appearance. For example, there was some form that all dogs had in common, *dogness*, and by virtue of this they were dogs, that is members of *Canis familiaris*. On this view there are absolute discontinuities between species, and since species-determining forms were viewed as unchanging, evolution of new species from existing species was viewed as a conceptual impossibility.

Associated with this view of species were various *morphological* species concepts according to which species membership could be determined by reference to *shape* (especially the shapes of anatomical features) construed as a measure of form. This idea fell into disrepute first through the observation of *polytypic* species, in which individuals of a given species display a great deal of variation with respect to characteristics, especially morphological characteristics. Second, there was the observation of *sibling species*. That is, good and distinct species that were sometimes so similar as to show no obvious morphological discontinuities—implying that speciation can occur without change of form. A good example here is a type of frog that used to be known as *Rana pipiens* (we now speak of the *R. pipiens* complex). This was a standard frog-model in physiological research. But labs started getting anomalous results, and careful studies revealed that what had been thought of as one species was in fact at least fifteen similar species (Berlocher 1998, 8).

What is needed is a way of thinking about biological species that reflects the facts of evolution. Any good textbook in evolutionary biology (for example, Futuyma 1998 and Price 1996) will provide you with an introduction to modern thinking about species and speciation, but the following observations will be helpful. From the standpoint of modern evolutionary biology, species are individuals that exist in space and time. They come into existence with speciation events, while they exist they have geographic distributions, and they go out of existence with extinction events. So what are they? Evolutionary biologists interested in mammals and birds (organisms that reproduce sexually), formulated the *Biological Species Concept* (BSC) as a first attempt to deal with this issue.

The BSC was one of the early fruits of the new synthesis that gave rise to modern evolutionary biology. As formulated by Ernst Mayr, who is one of the architects of modern evolutionary biology:

> A species . . . is a group of interbreeding natural populations that is reproductively (genetically) isolated from other such groups because of physiological or behavioral barriers . . . Why are there species? Why do we not find in nature simply an unbroken continuum of similar or more widely diverging individuals, all in principle able to mate with one another? The study of hybrids provides the answer. If the parents are not in the same species (as in the case of horses and asses, for example), their offspring ("mules") will consist of hybrids that are usually more or less sterile and have reduced viability, at least in the second generation. Therefore there is a selective advantage to any mechanism that will favor the mating of individuals that are closely related (called conspecifics) and prevent mating among more distantly related individuals. This is achieved by the reproductive isolating mechanisms of species. A biological species is thus an institution for the protection of well-balanced, harmonious genotypes. (1997, 129)

In these terms, morphologically indistinguishable sibling species, along with species whose members display a great deal of morphological variation, count as distinct species because they are reproductively isolated from other such groups of interbreeding natural populations.

From the standpoint of the BSC, it is necessary to think of species in terms of populations. A species may consist of a single population or several geographically distributed populations. The integrity of a species is thought of as being maintained by gene flow, that is, the exchange of genes within and between populations constitutive of the species. Consequently, processes and mechanisms that result in cessation of gene flow between populations are capable of driving the speciation process.

The central idea here is that with the cessation of gene flow between populations constitutive of a given species, genetic differences between those populations can accumulate to the point at which they become so different as to be reproductively isolated from each other (either physiologically or behaviorally). For example, with the cessation of gene flow between two populations adapting to new environments, mutations (contributing to variation among the alleles circulating in those populations) and natural selection (favoring some alleles at the expense of others) will drive genetic divergence between populations by bringing about changes in the frequencies with which alleles are found in those respective populations. Eventually these genetic divergences become so great that populations once capable of interbreeding can no longer do so. At this point speciation has occurred.

Many mechanisms capable of driving the occurrence of speciation events can be devised and tested in the laboratory (see Rosenzweig, 1995, ch. 5). Typical experiments might involve short-lived organisms, such as fruit flies, which can be subjected to various forms of selection and tracked in real time for fifty or more generations. Disruptive selection often plays a role in these experiments by favoring individuals with extreme traits at the expense of individuals with average values for those traits. (For example, if the trait was height, disruptive selection might work in favor of very short and very tall individuals—they would reproduce—while individuals of average height would face a reproductive penalty. The result of such selection, over many generations, would be two populations, one made up of tall individuals, the other made up of short individuals.) In this regard Rosenzweig has recounted the following anecdote:

> Bruce Wallace once showed me a new species of *Drosophila* [a fruit fly] he and his graduate students produced in his laboratory at Cornell. It fed exclusively on human urine, a previously unexploited ecological opportunity for [fruit] flies. They forced the speciation with artificial disruptive selection. Unfortunately the species is now extinct. The demigods at Cornell tired of the novelty and the fly lost its niche. (1995, 105–106)

Which of the possible mechanisms (derived from theory and laboratory experiments) actually play roles in driving the speciation process in nature is a current matter of scientific inquiry, one requiring careful field observations.

Of particular interest in connection with the issue of actually observing the occurrence of speciation is the possibility of speciation through polyploidy (or genome duplication). As noted above, genome duplication is a mutational event. When it happens, the organism with the duplicated genome is reproductively isolated from its ancestors because it has twice the number of chromosomes. Speciation happening this way occurs in a single generation, and has been observed to do so. It is estimated that at least 30% of speciation in plants has involved polyploidy. Some plants can, of course, fertilize themselves, so being cut off from their ancestors and their ancestors' *other* descendents is not so important as it would be for mammals, and their ability to hybridize more viably than animals is also believed to be important (Li 1997, 395–396; Maynard Smith 2000, 207–209). Speciation occurring this way has been observed, and hence macroevolution, as well as microevolution, has been observed.

Recent research has shown a role for speciation through polyploidy in insects, amphibians, and reptiles. A good example concerns the tree frogs *Hyla chrysoscelis* and *Hyla versicolor* that are found in the United States. These tree frogs are identical in appearance and occupy the same range (they can be differentiated on the basis of their respective mating calls). *H. versicolor* has arisen from *H. chrysoscelis* as the result of genome duplication (see Espinoza and Noor 2002).

While the BSC is helpful in the study of species and speciation, it has known limitations, and these become clear as one moves away from mammals and birds. Some clearly recognizable species consist of organisms that reproduce asexually (examples can be found among bacteria where, even though different species may share genetic material, they do so in ways decoupled from reproduction), whereas other species (for example, many plant species) have members that hybridize readily and viably with members of other clear and distinct species. In the case of these hybridizing species, gene flow between species can be an important source of genetic variation for evolution within the species. For these cases, the BSC is not helpful at all. As Price has recently observed, "Many species do not have enough sex: they are parthenogenetic, self-fertilizing, cloning or otherwise do not meet the criterion of biparental sexual reproduction. . . . Many other species have too much sex: they are promiscuous beyond the bounds of species identity, forming genetically open systems" (1996, 69). How can we cope with this situation?

Either the BSC is not a species concept with general applicability, or we have been mistaken about what is to count as a species. Perhaps bacteria and hybridizing plants are not, contrary to appearances, good and true species after all. This is not a conclusion that many biologists find to be satisfactory. There is a now a growing consensus among evolutionary biologists that the BSC provides an incomplete understanding of the nature of species, and recent developments in evolutionary biology have taken this into account (Pigliucci 2003).

Notice that the strategy adopted by those who champion the BSC is to take causal processes that create and sustain some good and distinct examples of species (in this case, processes inhibiting gene flow between sexual populations) and then to formulate a species concept in terms of an important result of these processes (reproductive isolation). In order to get beyond the BSC we need to give due consideration to the causal processes that create and sustain asexual species, hybridizing species, and so on. Moreover, we need to characterize what species are, in a way that does not simply reflect an end result (say, reproductive isolation) of just one of these causal processes.

To accomplish this end, we need to see if the processes that create sexual nonhybridizing species, sexual hybridizing species, asexual species, and so on, though different in mechanism, nevertheless share some functional similarities. It may then be possible to formulate a general species concept in terms of one or more of these functional similarities so that the different mechanisms can be seen as distinct causal pathways to a common functional end.

This idea has recently been discussed under the heading of the cohesion species concept, or CSC. Alan Templeton, who first formulated the CSC, characterizes a biological species as, "the most inclusive population of individuals having the potential for phenotypic cohesion through intrinsic cohesion mechanisms" (1989, 12). What does this mean? The strategy is to adopt a general concept of what a species is, while giving fair consideration to the plurality of mechanisms—intrinsic cohesion mechanisms—by means of which they are brought about and sustained. This way of proceeding allows us to talk of biological species by focusing on species in functional terms as maximally cohesive units, while simultaneously refusing to reduce our conception of biological species to the consequences of a particular causal mechanism to this end (for example, reproductive isolation).

Intrinsic cohesion mechanisms include gene flow, stabilizing selection (where individuals whose phenotypes diverge too far from the norm for the population are penalized through natural selection), developmental constraints (while the phenotype of an organism reflects complex interactions between the genotype and the environment, so that one and the same genotype might give rise to distinct phenotypes if the environments encountered are sufficiently different, it is nevertheless true that many phenotypes are not accessible from a given genotypic starting point because there is no developmental pathway leading in that direction), and reproductive isolation. In any given species, one or more of these cohesion mechanisms may be at work, but it may also be the case that mechanisms at work in one species may not be at work in another. Stabilizing selection, for example, might maintain the cohesion of an asexual bacterial population, while gene flow and developmental constraints might be at work in a sexual population. Some sexual populations are reproductively isolated from other such populations, while others hybridize. And as Price has noted (1996, 69), even hybridizing species usually retain distinctive species characteristics, with the hybrid zones where the hybrids flourish typically being narrow.

Evolutionary biologists have thus come to realize that the natural discontinuities that constitute species differences are the results of complex dynamical processes involving a multiplicity of mechanisms. How, then, do species differ? And where do new forms or morphologies come from? The following comments seem to be in order.

It is sometimes said that there is a 99% genetic (base pair) similarity between humans and chimpanzees. Doesn't this make them fundamentally similar to us—humans in ape suits, perhaps? The issue is rather more complex than it might at first appear. First of all, a lot of our DNA is not expressed and has no known functional significance—the so-called junk DNA. Such DNA diverges between species at a constant rate, and differences and similarities with respect to the degree of this divergence may record little about differences and similarities between species but rather may merely convey information about the time since divergence. In the present case, all it may mean is that the line that leads to modern humans diverged from the line leading to modern chimpanzees about 7 to 10 million years ago (Lewontin 1995, 15–16). This is about the same span of time separating deer from giraffes. Nevertheless, if we are so similar to chimpanzees at the genetic level, we are also clearly different both morphologically and behaviorally. How could this be explained? To deal with this question, biologists have had to examine the evolution of organismal development, thereby bringing about a new revolution in the way we think of evolution.

Noting the enormous diversity of animal forms, Wilkins has recently posed the puzzle this way:

> If these visible differences are a faithful reflection of the underlying range of genetic architectures, then few generalizations will be possible, and the task of understanding this genetic diversity will be correspondingly large. It is possible, however, that the visible diversity of morphology and development is misleading as to what lies beneath. Might there not be some significant, but hidden, genetic identities that exist between these seemingly highly different forms? (2002, 128)

This question could not be answered until the molecular revolution had taken place and biologists had PCR (polymerase chain reaction) machines to clone genes from many different animal species. The answer that has since emerged is that underneath the enormous phenotypic diversity we see in animal species, there are some deeply rooted genetic identities—profound evidence, even for creatures as different as humans and sea urchins, of common evolutionary ancestry. We and they are twigs on different branches of the same tree of life. It will not go amiss to at least explain the basic ideas behind this revolution in evolutionary biology.

At the genetic level, a distinction has recently emerged between structural genes (whose protein products play many roles and functions in the body, especially with respect to the origin, support, and maintenance of its infrastructure) and regulator genes (whose products turn the structural genes on and off, thereby regulating the protein production process). In tandem with this distinction, the idea has also arisen that genes do not work in isolation but work together in complex, interconnected networks—in fact, the study of this phenomenon belongs to a new branch of biology known as *genomics* (Carroll, Grenier, and Weatherbee 2001; Davidson 2001).

Organisms exhibit something known as hierarchical organizational complexity. An organism is made of organs, and organs come from tissues, which are made of cells, which in turn contain intracellular structures, which are made of macromolecules. At each level of the hierarchy, there are complex relationships between systems characterized at that level. But there are also complex relationships between the various levels (one reason that organisms cannot simply be reduced to their genes). Students of genomics are interested in the interactive complexity of genetic switching networks, their implications for systems elsewhere in the biological hierarchy, and the influence of these systems, in turn, on the behavior of the genetic switching networks. Important aspects of the biological significance of species differences between organisms arise because of differences with respect to this particular kind of organized complexity.

In such interconnected genetic networks, a single mutation in a regulator gene could have very large effects, bringing about changes in large patterns of gene expression (Gerhart and Kirschner 1997, 586–592; Kauffman 1993, 412; Wilkins 2002, ch. 14). Another way to make this point is to consider not humans and chimpanzees but rather humans and insects. Over the last ten years, many genes (including the so-called homeobox or *Hox* genes) have been found to regulate similar developmental roles in animals as distantly related as mammals and insects. And developmental biologists have been confronted with a puzzle known as the *Hox* paradox:

> How can bodies as different as those of an insect and a mammal be patterned by the same developmental regulatory genes? Very few anatomical structures in arthropods and chordates can be traced back to a common ancestor with any confidence. Yet to a rough approximation, we humans share most of our developmental regulatory genes not only with flies, but also with such humble creatures as nematodes and such decidedly peculiar ones as sea urchins. (Wray 2001, 2256)

One approach to this paradox was to simply deny that distantly related animals were that different after all. However, it has since become clear that developmental regulatory genes have acquired new roles in both insect and mammal lineages since divergence from a common ancestor.

This has led to a new approach to the *Hox* paradox in which it is recognized that though developmental regulatory genes have been conserved—so that similar genes are found in distantly related organisms—their interactions are not. Theorists now contend that many of the changes we see in animal evolution are the result of rewiring developmental gene networks (Wray 2001, 2256). Thus Carroll, Grenier, and Weatherbee observe:

> The recurring theme among the diverse examples of evolutionary novelties . . . is the creative role played by evolutionary changes in gene regulation. The evolution of new regulatory linkages—between signaling pathways and target genes, transcriptional regulators and structural genes, and so on—has created new regulatory circuits that have shaped the development of myriad functionally important structures. These regulatory circuits also serve as the foundation of further diversification. (2001, 167–168)

There is a good sense, then, in which developmental biology is showing that diversity of body forms is in the details of the genetic interactions!

While the gradual fusing of insights in developmental biology with insights drawn from evolutionary biology contains much truly exciting science, it has relevance for our concerns about the argument from design, for early mechanical design arguments hinged crucially on theories about how development took place. Here, then, is an example of how a gap in our knowledge can be closed, and it will now be presented and discussed.

Genes and Developing Machines

Darwin's original theory of evolution laid down a powerful challenge to the claim that organisms were machines. Adaptations—the very features of organisms that seemed to cry out for an account in terms of deliberate intelligent design—could be accounted for in terms of the operation of natural processes, and especially natural selection. Biology after Darwin has continued to challenge the viability of mechanical conceptions of organisms—this time from the standpoint of reproduction and development.

By contrast, if we journey back in time, we discover that mechanistically minded biologists of the seventeenth and eighteenth centuries had to explain the apparent generation and development of new organisms. How could one machine, the mother, give rise to other machines, the offspring? Mechanistic biologists formulated the theory of *preformationism* as an answer to this question. According to this theory, organisms are fully formed and differentiated in the seeds from which they are derived, with the developmental process being viewed as a process by which the preformed, miniature organism simply increases in size. In the context of human reproduction, an initial little person expands into a bigger person, who is finally given birth. And the little person is there literally as a little, preformed *person*, from the beginning of the reproductive process. No wonder there were moral strictures concerning abortion with this view of organismal development.

There were two schools of preformationism. One, led by Jan Swammerdam (1637–1680) held that individuals were preformed in the egg. He argued that an egg contained all future generations as preformed miniatures—a bit like Russian dolls, with one doll inside another, and so on. Another school, based on the work of van Leeuenhoek and Nicolas Hartsoecker (1656–1725) saw the preformed humans (or homunculi) as residing in sperm.

The mechanists saw organisms as machines but could not see how mechanical principles involving matter in motion could explain reproduction. Preformationism sidesteps the issue by seeing organisms as fully formed in their seeds, with all future generations of each species being preformed in miniature, one within another, at the time of initial design and creation by God. As Albrecht von Haller (1707–1777) put it: "The ovary of an ancestress will contain not only her daughter, but also her granddaughter, her great granddaughter, and her great-great granddaughter, and if it is once proved that an ovary can contain many generations, there is no absurdity in saying that it contains them all" (quoted in Mason 1962, 367). The preformationist school in effect solves the problems of reproduction and development by denying that reproduction occurs (future generations are already there in miniature) and by conceiving of development as an expansion of a preformed individual. Needless to say, modern biology has found no evidence of preformed individuals in either sperm or egg. Nevertheless, other options are possible for those who wish to see organisms as machines.

Paley, whom we met in the last chapter, thought of organisms as intelligently designed systems to be understood through an analogy with machines such as pocket watches. But watches, unlike organisms, do not reproduce and develop. Paley anticipated this objection as follows:

> Suppose, in the next place, that the person who found the watch should after some time discover that, in addition to the properties which he had hitherto observed in it, it possessed the unexpected property of producing in the course of its movement another watch like itself—the thing is conceivable; that it contained within it a mechanism, a system of parts—a mould, for instance, or a complex adjustment of lathes, files and other tools—evidently and separately calculated for this purpose. (1850, 14)

In Paley's self-replicating machine, it is imagined that the machine has a mechanical program and equipment to first manufacture the components of a watch and, second, to assemble these parts into a new, functioning, offspring watch, which inherits the ability to replicate itself from the parent watch. Paley's theory has the defect that while it offers an explanation of reproduction—it does not sidestep the issue as the preformationists did—it makes development mysterious. Mammalian parents, after all, do not make fully grown copies of themselves. It is as though big clocks make little pocket watches that somehow turn into big clocks.

In fact, it turns out that animal development is not very much like machine assembly at all. Development does not proceed through the initial fashioning of parts and subsequent assembly of those parts by the craftsman or even, in the Paley case, by the parent machine. It is actually a self-organizing process far more intriguing than a machine assembly process.

In humans, for example, development proceeds from the fusion of sperm and egg to form a zygote or fertilized egg. This process typically requires an appropriate maternal environment, but the mother does not deliberately bring about this fusion as a watchmaker (or self-replicating watch) might join two components together, nor is there a little person present, simply waiting for expansion to proceed until birth. The zygote undergoes mitosis, giving rise to two daughter cells, each having a nucleus containing the same number and kind of chromosomes as the cell from which they are derived. These cells in turn continue to divide and form a blastula—a relatively hollow ball of cells. The blastula stage is followed by the gastrula stage of development, characterized by the production of germ layers—layers of cells from which the animal's organs will be derived in the course of developmental time.

The important point is that all the cells in the developing embryo are genetically identical, and the question naturally arises as to how cells become specialized into liver cells, brain cells, kidney cells, and so on. They are not preformed in miniature. Moreover, it does not appear that the parent deliberately fashions differently specialized cells and then assembles them into an organism in the workshop of the womb.

We now believe that the process of cell differentiation depends on different genes being active in different cells. Structural genes (genes that make the proteins constitutive of the developing body's infrastructure) get turned on or off by proteins made by regulator genes (and there can be complex cascades of switching activity). Regulator genes are turned on and off in complex ways by chemicals in their environments. Different cell types result from different patterns of switching activity. As Maynard Smith and Szathmáry have recently noted: "In the cells of multicellular animals and plants, genes tend to have many different regulatory sequences, and are affected by many regulatory genes. Hence the activity of a particular gene, in a particular cell, can be under both positive and negative control from different sources, and can depend on the stage of development and of the cell cycle, on the cell's tissue type, on its immediate neighbors, and so on" (1999, 113). The developing embryo thus makes cells that, with appropriate environmental cues, self-organize into specialized cells and tissues. They do not require either preformation in miniature or an external guiding hand to account for their origin.

One mechanism by which specialization can occur is called *embryonic induction*, which Maynard Smith and Szathmáry explain through the following example:

> The lens of the vertebrate eye is formed by the differentiation of typical epithelial cells. What makes these cells different from other epithelial cells is that they come into contact with the eye cup, an outgrowth of the developing brain that will become the retina and the optic nerve. Thus a group of cells that would otherwise have become a normal component of the skin are induced to form a lens by contact with the eye cup. This has the desirable consequence that the lens forms exactly in front of the retina. (1999, 117–118)

At this point we are a long way from parental watches assembling offspring watches. The embryo develops as the result of its genes, its complex interactions with its environment, and its subsequent modifications of its local environment, including itself. In other words, there are complex processes of self-organization occurring in a developing system that has complex exchanges with its surroundings. But if this is how parts of the eye develop, how did the eye evolve? After all, Newton and Paley both cited the eye as a structure that called out for intelligent design.

The Eyes Have It

In Paley's exposition of the argument from design, he pointed to the human eye. He compared this with a pocket watch. Both are complicated, and the eye, like the watch, appears to have many finely crafted moving parts. It is beyond belief that such a complex, functional structure as an eye could have assembled just by chance; though a watch is much less complex, it would also be beyond belief if the watch were to assemble simply by chance. The watch needs a watchmaker to intelligently design and assemble it. The eye, too, needs a designer—a highly intelligent one—to explain its adaptive, functional features. Eyes are designed to see, as watches are designed to tell the time.

By contrast, if Darwin is right, the eye is indeed an adaptive, functional structure. But for Darwin, the eye did not arise by chance, nor was it the fruit of intelligent design. Eyes, being complexes or clusters of adaptations, must therefore be the fruit of the operation of the natural, unguided causal mechanism of natural selection. Darwin and Paley both agree that you do not get eyes just by chance. For Paley, they result from intelligent design, whereas for Darwin, they result from the operation of natural selection. But Paley does not tell us exactly how eyes were designed. Darwin does not tell us exactly how eyes resulted from selection. So do we then have no clear winner?

Not quite. Barring revelation, the way the eye (and everything else) was intelligently designed must remain a mystery known only unto God. We have no way to formulate or test hypotheses about intelligent design. We have no way to ask God, the way we could ask a watchmaker, exactly how it was done. By contrast, Darwin could point to evidence of the operation of natural selection with respect to numerous other structures and processes in humans and other species, all of which are as opaque as the eye from the standpoint of the intelligent design hypothesis. Still, these other structures are not eyes, and we must recall that Darwin, like Copernicus, took only the first steps. How have the competing explanations of the origin of the eye fared since the nineteenth century?

It is a sad fact that claims about the intelligent design of the eye remain as mysterious, unexplained, and undeveloped today as they were in Paley's day more than two centuries ago. By contrast, evolutionary biologists have discovered much about the evolution of the eye since 1859. While we do not currently have all the answers (and we will not find "all the answers" in any branch of science, all of which are works in progress), the fact that our knowledge and understanding has grown considerably with time marks an enormous difference in explanatory power between the static and empty design hypothesis and the dynamic and increasingly fruitful evolutionary hypothesis.

First, of the more than thirty animal phyla, about a third have species with proper eyes, a third have species with light-sensitive organs, and the remainder have no obvious means to detect light (Land and Nilsson 2002, 4). Comparative studies of extant animals reveal a nearly continuous range of intermediate cases with respect to sophistication of visual apparatus between, say, humans and earthworms. This brings out the evolutionary importance of the observation that distinct lineages descend from common ancestors with differing degrees of evolutionary modification. The effects of evolution are not everywhere the same. For every species in its ecological context, there are limits to how much visual information it can use. The evolution of earthworms has been such that they do not require human eyes to make a successful living and do not have the nervous systems to process the information that those elaborate structures can provide. Humans have evolved (and have the nervous systems to use) something more sophisticated than the simple light sensitivity of earthworms.

The range of structures we see in living species today thus conveys valuable information about the various evolutionary gradations that occurred (and are certainly possible, since we actually see them) over time, as modern eyes such as ours evolved by degrees from the simpler structures possessed by ancestral species and ultimately back to species with a single, light-sensitive cell. The fossil evidence hints at an origin of the first eyes about 530 million years ago.

A single light-sensitive cell is better than nothing in the land of the blind. There is its selective value. More than one such cell confers an advantage, too, if only through redundancy and insurance against loss. Directional vision would be even better, and it can be achieved by shielding the light-sensitive cells with a pigment. As noted by Land and Nilsson, there are two ways of proceeding at this point—two distinct evolutionary trajectories that can be taken:

> Either more photoreceptors are added to exploit the same pigment shield, or the visual organ is multiplied in its entirety. The two alternatives lead to simple (single chambered) and compound eyes respectively. . . . During the early stages of eye evolution there would be little difference between the efficiency of the two solutions—single chambered or compound. . . . Irrespective of whether evolution originally

> takes the path towards a simple or compound eye, shielding will soon turn out to be an inefficient mechanism on its own. As the spatial resolution is improved by adding more picture elements, the directionality of each photoreceptor will need to improve as well. It is at this stage in eye evolution where more elaborate optics, in the form of lenses or mirrors, will significantly improve the design. Because even the slightest degree of focusing is better than none at all, lenses or mirrors can be introduced gradually, with a continuous improvement in performance. (2002, 7)

Are there clues in extant species for how this process could have occurred, yielding more sophisticated single-chambered eyes with lenses such as we enjoy?

It would be useful—that is, confer a selective advantage—to be able to differentiate lighter from darker regions of the environment. A simple way to achieve this—seen in the limpet *Patella*—is to have a **V**-shaped pigmented pit lined with light-sensitive cells. Pits—essentially depressions in tissue—are easy enough to make, and piteyes are fairly common. And as Land and Nilsson go on to observe:

> In many gastropods, the abalone *Haliotis* for example, the mouth of the pit is drawn in to give the eye a more spherical shape, and a narrower opening, restricting the acceptance angle to perhaps 10 degrees. While this results in an improvement in the eye's resolution, it is obvious that to pursue this line any further will produce eyes in which less and less light reaches the image. Thus this is not a particularly good evolutionary route to follow. The only animal to have pursued this to its logical conclusion is the ancient cephalopod mollusc *Nautilus*. A much better solution is to evolve a lens. In the snail *Helix* this is simply a ball of jelly which converges the light rays a little, though not enough to form a sharp image. However in the periwinkle *Littorina*, and many other gastropod molluscs, the lens has evolved into a sophisticated structure with a graded refractive index, and excellent image-forming capabilities. (2002, 56–57)

Thus, as light-sensitive cells are better than nothing in the land of the blind and open up new ways to make a living and specialize in the economy of nature, so pigmented pits are better than pigmented surfaces; pits that have a narrow opening can give eyes like pinhole cameras. Better still is some degree of narrowing and a ball of jelly. For in the land of unfocused light, some focusing is better than none. New ways of making a living accompany these innovations, and structures once there can always be improved by natural selection.

Once there is focusing, niches can open up in which more is better than less, and so on.

The point is made if we realize that to evolve a human eye, we do not need everything at once, and more rudimentary structures of varying degrees of complexity can get a job done that is selectively advantageous in the environment in which it is found. Small wonder now that this is known about the eye that creationists have had to find other gaps in which to insert their intelligent designer. As we shall see shortly, rather than argue about anatomical structures such as eyes, intelligent design theorists have gone hunting for new gaps—gaps that should stand to the modern evolutionary biologist the way the eye stood to Darwin. Biology has recently undergone a molecular revolution, in which the focus of biological inquiry has shifted from large structures such as eyes to structures and processes within our cells.

In the case of eyes, these molecular studies have yielded some intriguing surprises. Structures as morphologically different as insect eyes and human eyes share important similarities at the genetic level, and research has focused on a regulator gene known as *Pax6*. This gene has been shown to play a crucial role in eye development in vertebrates and invertebrates. Mutations in *Pax6* result in similar developmental defects in human, mouse, and fruit fly eyes (Gerhart and Kirschner 1997, 33–34). Moreover, *Pax6* from a mouse has been shown to promote fruit fly eye development in ways characteristic of fruit flies. This is evidence of conservation of function in widely separated evolutionary lineages and hence descent from common ancestors. The eye, far from being a challenge to evolution, has turned out to be a vindication.

Intelligent designers are nothing if not persistent, however, and have followed the molecular pioneers to try to exploit the gaps in our knowledge that are typically present when pioneers enter virgin territory. We will shortly see that, as with the case of vision, there is less to these new design arguments than meets the eye.

3

Thermodynamics and the Origins of Order

The very existence of organisms of any kind involves the existence of complex, structured, highly organized, and ordered states of matter. Organisms are many orders of magnitude more complex than pocket watches. It is very natural to want an explanation of how such orderly, organized, complex states of matter could come to exist. Curiosity about these matters has led biologists, chemists, and physicists to consider some of the deepest and most fundamental laws in modern science: the laws of thermodynamics. But real scientists are not the only ones interested in the laws of thermodynamics. These same laws have also attracted the attention of creation scientists who think that these laws forbid the very appearance of complex, organized structures as the result of the operation of natural, unguided, causal processes. According to these folk, the complex order and organization we see in nature must result instead from intelligent design and supernatural causation.

Because these latter sorts of claims, trumpeted long enough, are apt to gain some credibility, this is yet another Augean stable that must be cleansed. But in this cleansing process we will derive much intellectual satisfaction from the discovery that, far from forbidding the appearance of complexity and organization, the laws of thermodynamics provide the basis for an understanding of these curious phenomena. Our journey in this chapter will take us into the strange territory of the self-organizing, self-assembling properties of physical systems driven by flows of energy.

Anyone who has taken undergraduate physics knows that thermodynamics is a tricky subject that involves a fair bit of subtle mathematics. I am not going to give a mathematics lesson here. I will leave that to Peter Atkins, whose book, *The Second Law: Energy, Chaos and Form* (1994), is a very fine exposition of thermodynamical principles in ways accessible to a curious nonspecialist. Instead, I will try to convey such fragments of thermodynamics as are needed to understand the controversies concerning intelligent design. However, before looking at some real science, we must first examine some pseudoscience, alas.

Creation Science and the Second Law of Thermodynamics

Henry Morris has led the creationist charge against evolution through the invocation of the laws of thermodynamics. In his book, *The Troubled Waters of Evolution*, he argued: "Evolutionists have fostered the strange belief that everything is involved in a process of progress, from chaotic particles billions of years ago all the way up to complex people today. The fact is, the most certain laws of science state that the real processes of nature do not make things go uphill, but downhill. Evolution is impossible" (1974, 110). Later he added: "There is . . . firm evidence that evolution never could take place. *The law of increasing entropy* results in an impenetrable barrier which no evolutionary mechanism yet suggested has ever been able to overcome. Evolution and entropy are opposing and mutually exclusive concepts. If the entropy principle is really a universal law, then evolution must be impossible" (1974, 111).

Let's be clear about this: If Morris is right, the issue is not just biological evolution, for organismal development, going "uphill" from fertilized egg to adult, must be impossible, too. All processes in unguided nature, if Morris is right, are processes by which things inexorably run down, break down, decay, and go "downhill."

Organismal development, like evolution, happens. Does this mean that the laws of thermodynamics are in error? Does the universe need intelligent guidance in the form of supernatural causes to stop the inexorable downhill trend? No. The problem lies with Morris's failure to think carefully about thermodynamics. Of course, Morris represents the older tradition of young Earth creation science that prominent intelligent design theorists wish to repudiate. But misunderstandings about these matters have found more artful proponents in the context of intelligent design theory. One such is William Dembski, and we will meet him again at the end of this chapter.

A Tale of Two Laws

Thermodynamics began with the study of the relationships between heat and work. Interest in these matters arose in the context of the steam-powered technologies that were crucial in the industrial revolution. While steam engines (and the modern fruits of the industrial revolution such as air conditioners, refrigerators, and heat pumps) are examples of thermodynamical systems, the resulting laws of thermodynamics apply to all physical systems, be they of interest to the physicist, the chemist, or the biologist.

At an intuitive level, a physical system is an arrangement of physical objects with a boundary that separates it from other such systems. Boundaries may sometimes be complex, even a bit blurred, but the fact that we can differentiate refrigerators from hair dryers, hurricanes from the rest of the weather system, the sun from the planets in orbit around it, and so on, at least suggests that we have an eye for physical systems. There are two distinct types of systems, and we need to be clear about what they are.

First, there are isolated systems sometimes referred to as closed systems. These are systems in which neither matter nor energy can be transported across the boundary of the system. In textbooks, it is often convenient to talk of isolated systems, for though they cannot literally be found in the real world, they are nevertheless idealizations of real systems that permit simplified explanations of tricky principles. (Similarly, in Newtonian gravitational theory, physicists, for reasons of convenience, sometimes think of planets as masses located at points in space.) We will shortly meet physical systems called *heat engines* that are closed or isolated in this sense. Second, there are open systems. These are systems that have exchanges of matter and energy with their surroundings. Organisms and their cells are open systems in this sense, as are hurricanes, river systems, and tornadoes.

The First Law of Thermodynamics is known as the *law of conservation of energy*. Intuitively, it says you cannot get something for nothing. Slightly more technically, it says energy can be neither created nor destroyed, though it can change its form and the way it is distributed. More technically still, it says that the energy in an isolated system remains constant over time. Consider an isolated system in the form of a box initially containing air at room temperature and a lump of red-hot iron. Over time, the iron cools, and the surrounding air in the box warms, until equilibrium is reached, at which point the iron and the air are at the same temperature. The total energy in our imagined system does not change over time, but the distribution of the energy has clearly changed. Energy is even permitted to change its form—when a candle burns, chemical energy in the candle is transformed into thermal energy—just so long as you do not get something from nothing.

The Second Law of Thermodynamics builds on these intuitive insights. Though the energy in an isolated system is constant over time, the energy that can be used to do work—that is, to run a machine or drive other physical processes, such as chemical reactions—undergoes changes. In particular, the amount of usable energy—that is, energy available for work—tends to a minimum. Intuitively, our red-hot lump of iron discussed before is a heat source that radiates its thermal energy into its surroundings—heat flows from the hotter to the cooler—until equilibrium is reached. Until equilibrium is attained, the iron is an energy source capable of driving physical processes in the box by virtue of the energy difference that exists between it and its surroundings. At equilibrium, except for random fluctuations, there is no energy difference, and no work gets done.

At an intuitive level, a car runs because some of its parts are much hotter than their surroundings as the result of the conversion of chemical energy in the fuel to heat energy in the cylinder, which makes gases expand, which in turn drives the pistons in the cylinders up and down. Heat is vented to the environment by hot gases leaving the exhaust pipe and through radiation and convection, notably from the engine block and exhaust manifold. When the fuel is exhausted, the engine stops, and the car gradually settles into a state of thermal equilibrium with its surroundings.

An intuitive statement of the Second Law says that whenever you have only a fixed quantity of energy, you cannot use all that energy to do work. An energy source at equilibrium with its surroundings still contains energy but not the kind available for doing work. (Our lump of iron may still be warm at equilibrium, but with no temperature difference between it and its surroundings, it can no longer drive physical processes in the box.) A more technical view of the Second Law will say that in an isolated system, the entropy of the system tends to a maximum and the energy available for work tends to a minimum. But this technical statement involves reference to a new physical quantity, called *entropy*, which, unlike temperature, is not something we talk about in everyday life.

What is entropy? It is a term that has been subject to much abuse by creation scientists and by others who have found it necessary to appeal to the laws of thermodynamics in popular publications. We are sometimes told that increasing entropy results in increasing disorder, thereby linking entropy to the idea that it somehow corrupts order. But *order* and *disorder* are terms that have anthropocentric overtones, like *tidy* and *untidy*, and are thus not well suited to a discussion of basic physical laws, which care nothing for the fastidiousness of people and the condition of their belongings and other surroundings.

What we need to do is consider a simple physical system that consists just of an energy source leaking energy to an energy sink, and between the source and the sink we will have some physical objects through which the energy must flow on its way to the sink. The situation envisioned here is diagrammed in figure 3-1.

Let us denote the temperature of the heat source by T_1 and that of the heat sink by T_2, and let's assume that initially the system is such that $T_1 > T_2$, so there is a temperature difference between the source and the sink. Assume also that a quantity of heat denoted as ΔQ_1

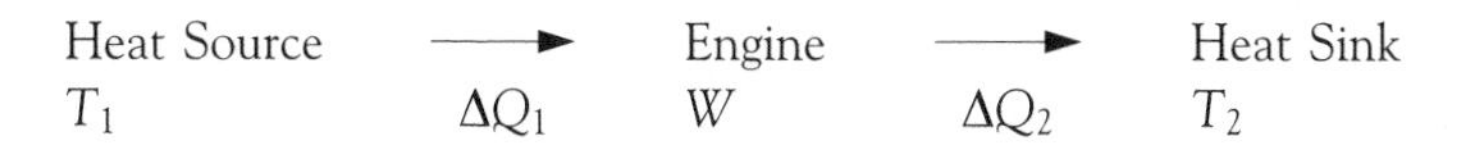

Figure 3-1. Schema for a heat engine.

flows from the source to the engine. Suppose as a consequence of this heat flow that the engine does work W and in the process dumps a quantity of heat ΔQ_2 into the sink. An engine need not be a machine; it is simply a physical system that does work as energy flows through it. Work is often done in cycles; for instance, a piston in a cylinder cycles by going up and down. A water wheel partially immersed in a flowing stream cycles by rotating round and round. Work is also done in the cells of your body during metabolic cycles. Work is a measure of energy transformation. In a process involving work, energy gets redistributed.

With all this in mind we can say, in accord with the First Law, that the amount of work must be calculated as:

$$W = \Delta Q_1 - \Delta Q_2 \qquad [1]$$

Thus, in getting work we didn't get something for nothing. Our heat source contains thermal energy available for doing work. Some thermal energy left the source. Work was done. In the process, a smaller amount of thermal energy was dumped into the heat sink. Work done is equal to the difference between the two quantities of thermal energy. Thus energy has been redistributed, but it has not been created or destroyed.

Because of redistribution, less energy is now available for doing work—the amount of usable energy has decreased. Physicists describe this situation by saying that the entropy of the system has increased. The change in entropy, ΔS, of a system is defined in terms of the heat, ΔQ, supplied to the system divided by the temperature, T, of the system:

$$\Delta S = \Delta Q / T \qquad [2]$$

Peter Atkins observes of this simple equation:

> If energy is supplied by heating a system, then *Heat supplied* is positive (that is, the entropy increases). Conversely, if the energy leaks away as heat to the surroundings, *Heat supplied* is negative, and so the entropy decreases. If energy is supplied as work and not as heat, then the *Heat supplied* is zero, and the entropy remains the same. If the heating takes place at high temperature, then *Temperature* has a large value; so for a given amount of heating, the change of entropy is small. If the heating takes place at cold temperatures, then *Temperature* has a small value;

> so for the same amount of heating, the change of entropy is large. (1994, 34)

There are various qualifications—we have ignored friction, assumed large sources and sinks, and so on—but this is a good start to the study of entropy.

As applied to our heat-driven engine, heat leaks away from the source so the entropy ΔS_1 of the heat source decreases thus:

$$\Delta S_1 = \Delta Q_1 / T_1 \tag{3}$$

But heat is supplied to the sink, so the entropy ΔS_2 of the heat sink increases:

$$\Delta S_2 = \Delta Q_2 / T_2 \tag{4}$$

Because $T_1 > T_2$, the decrease in entropy of the source ΔS_1 is less than the increase of the entropy of the sink ΔS_2, so the total entropy change ΔS is calculated as:

$$\Delta S = \Delta S_1 + \Delta S_2 \geq 0, \tag{5}$$

and entropy for the whole system increases. But the magnitude of the increase gets smaller and smaller as T_1 decreases by losing heat and T_2 increases by gaining it, and so, in accord with the Second Law, the entropy of the whole system tends to a maximum as time goes by, and the source and sink get closer to a state of equilibrium where $T_1 = T_2$.

One thing that emerges from this brief study of entropy is that while the total amount of energy remains constant in our system, it is subject to redistribution in such a way that less and less is available for work. The quantity of energy is constant, but its usefulness or quality, measured as availability for work, is not. A failure to understand this point is of great practical importance, because as Atkins has noted:

> As technological society ever more vigorously burns its resources, so the entropy of the universe inexorably increases, and the quality of the energy it stores concomitantly declines. We are not in the midst of an energy crisis: we are on the threshold of an entropy crisis. Modern civilization is living off the corruption of the stores of energy in the universe. What we need to do is not conserve energy, for Nature does that automatically, but to husband its quality. In other words we have

> to find ways of furthering our civilization with a lower production of entropy: the conservation of quality is the essence of the problem and our duty toward the future. (1994, 39)

It goes without saying that our planetary system is being warmed from space by a large heat source called the Sun, and as energy flows in complex ways on and into the planet, before being reradiated back into cold space, physical systems, including ourselves, do all sorts of interesting work.

Because of the Sun, the Earth is far from a state of thermodynamical equilibrium with its cold surroundings. The hot core of the planet helps, too, by driving physical processes on a planetary scale (for example, volcanoes and continental drift). These processes, over long periods of time, have shaped the geography and geology of the world we live in. With all this in mind, we now need to examine how thermodynamical issues have arisen in debates about evolution.

Thermodynamics and the Origins of Order

In the last section, we saw that when dealing with a complex system consisting of sources, sinks, and engines, entropy calculations had to look at the entropies of the parts and see how they contributed to the entropy of the whole. Thus, in the equation $\Delta S = \Delta S_1 + \Delta S_2 \geq 0$, we saw that the net entropy of the total system increases as required by the Second Law, despite the fact that the entropy ΔS_1 of one of the parts, the heat source, nevertheless decreased. The point here is that the mandated increases in net entropy that are required by the Second Law are completely consistent with localized decreases in entropy. However, to understand the implications of this observation, we must look even closer at the meaning of the Second Law.

One of the great achievements of physics in the late nineteenth century was the forging of connections between the basic ideas of thermodynamics and basic ideas from atomic theory, according to which the familiar objects of everyday experience are vast conglomerations of molecules, and ultimately atoms—tiny, microscopic physical systems in complex states of jostling motion. Physicists at this time thought of atoms as billiard balls writ small, and so, for the sake of simplicity, shall we. And now, thinking along these lines, we have to reexamine some of the basic ideas of thermodynamics.

In our discussion of heat engines in the last section, we saw that heat tended to disperse and, in particular, to flow from the hotter to the cooler. One way to think about the Second Law is to see it as saying that in isolated systems, energy (for example, thermal energy) tends to disperse. Putting things this way invites the question of how energy disperses. Part of the answer is that macroscopic systems like lumps of iron are made of atoms and molecules. Atoms and molecules carry energy as a result of their motion (whether vibratory or translational). Energy is dispersed when atoms and molecules change their locations by moving about in space or when they transfer it to other atoms and molecules by bumping and jostling each other. The hotter macroscopic systems are, the more energy their atoms and molecules have, and hence the more vigorously these atoms and molecules move, vibrate, and jostle. This is the basic insight behind the kinetic theory of heat.

According to the kinetic theory of heat, what we experience as heat is due to the motions of atoms and molecules—the faster they move or vibrate (on average), the hotter the systems containing them. You can try a simple experiment in applied thermodynamics. Take a wire coat hanger and bend the wire back and forth several times. You are doing work to the coat hanger in this process. The place where the wire has been bent back and forth will become hot in this process, and you can feel the heat. In this case, mechanical energy has been converted into heat energy. The atoms in the coat hanger where it is hot are now moving faster than they were before you added energy to them.

In the heat engine we discussed in the last section, some thermal energy was redistributed and work was done—pistons may have gone up and down in cylinders, wheels may have rotated, and so on. These are useful motions of matter, and we exploit such motions every day of our lives. But these useful motions of matter reflect properties of the motions of the atoms and molecules out of which the engines are made. To better understand this last point, we must differentiate between coherent motions of atoms and random, incoherent, thermal motions of atoms.

When a piston in a cylinder goes up and down, there is a net upward and downward motion of the atoms making up the piston. These are coherent motions. When we get work from a system, it is because we are able to use energy to induce and sustain coherent motions of the atoms making up our machine. Consider a car: Coherent motions in one part (the reciprocating motion of the pistons in the engine block) are converted through coherent motions in other parts (cranks and gears) into coherent, rotary motions of the wheels. By virtue of this coherent motion, by burning gas, you can transport yourself from place to place.

Alternatively, you could use energy simply to make a system hot. Your stove takes chemical energy in gas (or electrical energy) and converts it into heat energy. When gas burns (combines with oxygen), energy disperses through the incoherent, random motions of molecules. These molecules jostle molecules in the pan on the stove, which disperse energy by transferring it to the water molecules in the pan. As these jostle faster, the water gets hot, and you can make tea.

The real thermodynamical systems we encounter—cars, for example—involve a combination of coherent and incoherent motions of atoms and molecules. A sensible car owner tries to reduce the unnecessary induction of incoherent thermal motions in her car by making sure that it is properly lubricated to reduce friction, which is a well-known source of heat. Some parts of the car get very hot—for example, the spark plugs—and these wear out faster than other parts. In this last automotive observation, we can begin to tie in the concept of entropy to those of order and disorder.

Increases in entropy in a system result from increases in the incoherent, random, thermal motions of atoms and molecules making up the system. Decreases in entropy result from reductions in such incoherent motions. Let's now go back to our latest version of the Second Law, according to which energy tends to disperse. Peter Atkins has observed:

> The concept of dispersal must take into account the fact that in thermodynamic systems the coherence of the motion and the location of the particles is an essential and distinctive feature. *We have to interpret the dispersal of energy to include not only its spatial dispersal over the atoms of the universe, but the destruction of coherence too.* Then *energy tends to disperse* captures the foundations of the Second Law. (1994, 62)

Energy can be dispersed by one atom transferring energy to another or when the atom carrying the energy changes its location. A car, for example, requires coherent motions in many of its parts, but it also requires that parts, made of atoms, don't get shaken off, thereby changing their locations in space in such a way as to destroy the structural coherence of the car. Energy is redistributed in the car factory, work is performed, and this gives you an organized structure like a car, but as time goes by, energy disperses, and despite the best efforts of mechanics, all cars will eventually degrade in the process.

Let's ignore organisms for the moment and think carefully about nonliving physical systems—we will call them *dynamical structures*—that come into and go out of existence near the surface of the planet as the result of energy flowing in from the sun, to be radiated back out to space. A hurricane is a good example. There is a season when these systems are spawned. They are fed by energy from the Sun that has been absorbed by oceans. Hurricanes can exist for a week or more, and they are visible from outer space as rotating spiral patterns, sometimes up to 1,500 km across and 15 km high. (Tornadoes are more localized structures that exist on time scales of minutes.) Both hurricanes and tornadoes involve the emergence of coherent motions of matter on large scales. They are not just random winds; they are highly organized systems. This is why we can discern predictable spiral and funnel shapes. Hurricanes are dynamical structures precisely because they exist due to the energy-driven, coordinated, coherent motions of large quantities of matter. Hurricanes do work and actually have an enormous power output that may be as much as 10^{13} watts.

To get a hurricane, you need several things to come together, as natural mechanisms operate in accord with unintelligent natural laws during hurricane season. The ocean must be at least twenty-seven degrees Celsius. You need a latitude of at least five degrees north or south of the equator. You also need a region of low pressure at sea level. Hurricanes, though undesigned, behave like the heat engine discussed in the last section. The ocean (warmed by the Sun) is the heat source. There is a temperature difference between the surface of the ocean and the upper atmosphere (where it is much colder)—and hurricane intensity reflects this temperature difference (as well as other factors, such as pressure differences).

Here is how the hurricane forms without the help of intelligent design. A region of low pressure draws in moist air from surroundings at higher pressure. This causes moisture in the air to condense, and the water, in changing state from gas to liquid, releases thermal energy as latent heat. The resulting warm air rises, drawing in more air from below, and hence from outside the immediate region of the forming hurricane. The water in this air condenses as well, releasing more heat. In this way, the eye of the storm forms, and the hurricane becomes a self-sustaining structure whose existence depends on energy flowing through the system and getting redistributed in the process. The spiral patterns shown by hurricanes result from the operation of Coriolis forces, which do not exist at the equator (hence the need for a latitude of at least plus or minus five degrees).

Coriolis forces are *apparent* forces arising because the earth is rotating on its north-south axis; that is, the affected objects appear to move as if they are being acted on by a force. In the Northern Hemisphere, instead of moving in a straight line—say, from north to south—the affected objects deflect to the west. In reality, the effect is due to the earth's rotational motion. If you look down from the North Pole, the rotation is counterclockwise. A point fixed to the equator is actually moving at around 1,100 km per hour; a point at the North Pole is not moving at all. Particles, such as those in the air or clouds, that are not fixed to the moving surface of the Earth tend to deflect to the west as they move to regions of low pressure. They thus give the appearance of being acted on by a force. The result is that, in the Northern Hemisphere, air circulates in a counterclockwise direction around a region of low pressure. You can see this on weather maps in newspapers and on television.

Moist air, then, is drawn to the center of the hurricane, is warmed through the release of heat energy, and then ascends the wall of the eye. If there is no disruption from wind shear, the air cools at the top of its ascent, radiating heat to space, and is pumped horizontally to the extremities of the hurricane, descending as it cools, and the whole process thereby draws in more moist air at the bottom in order to perpetuate the hurricane. The net coherent motion of matter in the rotating structure is capable of doing work. Hurricane formation involves a localized reduction in entropy.

The entropy requirements of the Second Law are such that the orderly motions and structures in the hurricane must be more than offset by disorderly, incoherent motions elsewhere. And anyone living in a coastal community where a hurricane has made land-fall knows exactly what is meant by way of the incoherent motions of matter involved in settling the entropy accounts to satisfy the requirements of the Second Law. Landfall also disrupts the hurricane mechanism because of friction and a lack of moist air. In a sense, it is starved out of existence by getting cut off from the energy source that powers it.

In the hurricane, there is a localized decrease in entropy as structure and pattern emerge. But this decrease is more than offset by increases in the entropy of the environment with which the hurricane interacts. The Second Law is satisfied. Creationists are simply mistaken that all natural unguided processes must go *downhill*. Nevertheless, if the emergence of structure is consistent with the Second Law, important issues need to be discussed. In particular, what is it about physical objects that permits them to organize into structured, orderly, organized complex systems? This is an interesting topic that will take us into a discussion of the science of self-organization.

Some Secrets of Life

We cannot escape the Second Law; local reductions in entropy have to be compensated for by entropic increases elsewhere. But, to use Henry Morris's language (though not his sloppy reasoning), it is not simply that the Second Law permits things to go uphill, as long as it is compensated for elsewhere by other things going downhill. Rather, processes can be coupled so that as something goes downhill, it can make other things go uphill. To use an example from Atkins (1994, 167), consider a heavy weight *A* tied to a light weight *B* by a length of string. If the string goes over a pulley wheel and the heavy weight *A* is allowed to fall downhill, in the process of falling it will raise uphill the light weight *B* to which it is attached or coupled. All that is needed is gravitational energy. Once the light weight *B* is raised in this manner, suppose that the heavy weight *A* is replaced with a weight *C* that is lighter still than the weight *B* that has just been raised. Now the weight *C* can be raised while the weight *B* falls, and so on. In this way, interconnected sequences of uphill changes can occur, provided there is an overall downhill trend.

Luckily for all of us, chemical versions of this weight-lifting feat go on in our bodies all the time. There is no guiding intelligence, just chemical mechanisms operating in accord with the laws of chemistry. In our cells, the molecule that carries energy enabling our cells to do work to sustain themselves and the tissues and organs to which they belong (hence to sustain life) is ATP (adenosine triphosphate). ATP is needed for many different functions in a cell. In particular, lots of energy is needed to synthesize proteins from amino acids, and ATP provides the energy to drive the process of polymerization, whereby long, structured, organized, lower entropy protein molecules are assembled from smaller, less organized, higher entropy amino acid building blocks.

ATP is synthesized (made) from a simpler molecule called ADP (adenosine diphosphate) through the addition of a cluster of atoms known as a phosphate group (PO_4). The energy needed to attach this cluster of atoms (the heavy weight whose falling raises the light weight) comes from the oxidation of sugar—glucose ($C_6H_{12}O_6$)—which is transformed into simpler, less structured, higher entropy carbon dioxide (CO_2) and water (H_2O) in the process. Glucose in turn is a light weight raised by the heavy weight of carbohydrates going downhill as a result of digestive (metabolic) action.

(Carbohydrates themselves are long, structured molecules typically synthesized within organisms such as plants; they are made up of chemically linked chains of sugars. With the help of energy from the sun, photosynthesis enables plants to make glucose from carbon dioxide and water, and this can then be used to make carbohydrates. The metabolic breakdown of carbohydrates resulting in glucose molecules is a process involving increases in entropy through the production of smaller, less organized molecules.)

The energy acquired by ATP from the oxidization of glucose can be surrendered by removal of the phosphate group. This energy can be used to help forge a peptide bond between amino acids in the process of the metabolic synthesis of protein molecules. Proteins and carbohydrates, once synthesized, are both molecules that can be consumed by other organisms. To get hold of a source of protein, usable energy will have to be expended in tracking it down, and the environment will be heated, and so on. All this activity is consistent with satisfaction of the Second Law.

An important concept here is that of a pathway. Pathways are causal routes by means of which changes in the world occur in accord with mechanisms obeying scientific laws. If changes happened without rhyme or reason or pattern—if, in short, there were just uncaused, random happenings in which events were not tied as cause and effect—then there would be no need to consider pathways. But such is not our world. Pathways exist precisely because many changes in the world around us happen in accord with the operation of causal mechanisms of various kinds. Biochemical pathways are sequences of chemical reactions by which biochemical changes are effected.

We have just looked at some fragments of real pathways, and we shall examine more in later chapters. A simple pathway might be represented as a simple, linear sequence of reactions:

$$A \rightarrow B \rightarrow C \rightarrow D, \quad [6]$$

by means of which a substance, *A* (an initial substrate), is transformed into another substance, *D* (a final product). As long as there is usable energy flowing through the pathway, feed *A*s in and you get *D*s out! As noted previously, pathways may be linked, so that products of one pathway become the initial substrates of the next. There can also be "loops" of interconnected pathways in which the final product of a sequence of reactions can be used to feed in as the initial substrate to get the cycle going again. Such cyclical reactions are driven by usable energy as it flows into the cycle at various points and then exits with higher entropy. The Krebs (citric acid) cycle, central to the metabolism of aerobic organisms, is a good example of such a cycle.

There are also genetic pathways in which one gene activates another and so on. There are developmental pathways in which the development of an adult organism results from orderly sequences of developmental events, proceeding, for example, from those initiated by fertilization of an egg. All these events depend on energy flows in which energy is conserved but becomes less usable as its entropy increases.

We have just looked at mechanisms within organisms and hurricanes by means of which both are sustained. Though organisms are very different from hurricanes, especially with respect to size and complexity, they exhibit some important similarities from the standpoint of the science of thermodynamics. Both are examples of what is known as *open-dissipative* systems.

Open-dissipative systems have exchanges of matter and energy with the environment that surrounds them, and they exist only so long as energy flows through them. Such a system takes in energy available for work, work is done internally (possibly to sustain it, possibly to make it grow, possibly to make it reproduce), and it then dissipates waste back into the environment. We humans, for example, take in low-entropy (organized) food molecules (proteins, carbohydrates). Work is done internally. We excrete smaller, less organized, higher entropy molecules. We also dump heat into the environment.

But open-dissipative systems are also dynamical structures. They are not permanent features of the world. They come into existence, their internal dynamics and environmental interactions permit their existence typically for a finite time, and though resilient in the face of environmental perturbations, destructive perturbations can destabilize and destroy the internal dynamical order, coherence, and integrity of the system. Landfall will do in a hurricane, and there are many ways in which we humans can be fatally perturbed. Such systems thus go out of existence. They are thus temporary islands of order rising out of, persisting, and subsiding back into the increasingly incoherent universe. In essence, they are components of pathways by which the universe expends usable energy and increases entropy in the process. In the last section of this chapter and in chapter 6, we will return to issues about the entropy of the universe as a whole.

Back on earth, hurricanes have a lifetime of a week or more, tornadoes have life spans measured in minutes, and humans (with modern medicine) have an average lifetime that may be rather more than three score and ten years. The red eye of Jupiter, observed by Galileo and still observable today, is a stable open-dissipative structure in the Jovian atmosphere (similar in some ways to a hurricane) that has been around for centuries.

But how do open-dissipative systems form? The key is thermodynamical equilibrium. In the 1860s, the French physiologist Claude Bernard pointed out in connection with organisms that equilibrium was death. He was the first theorist to realize the importance of the internal environment of organisms—the *milieu intérieur*—and that life required the internal environment to be out of thermodynamic equilibrium with its surroundings. As scientists in the last half of the twentieth century realized the importance of studying the dynamics of what have come to be known as nonequilibrium systems, many discoveries have been made about the characteristics of these systems. We have discovered that collections of physical objects of many different types, when taken away from equilibrium as the result of flows of energy, will spontaneously interact, self-assemble, and self-organize into complex, ordered, organized, dynamical systems.

I have just suggested that energy flows taking physical systems away from equilibrium can result in the emergence of structured, organized states of matter. What does this mean? Structure and organization may be spatial or temporal. So structure might appear to us in the form of sequences of changes that occur over time, as in the Krebs cycle. It may also appear in the form of coherent, nonrandom arrangements of physical objects in space—for example, the atoms making up amino acids, which in turn are polymerized into lower entropy complex structures such as proteins. It may involve both, as when spatial organization changes over time, as it does in developing organisms.

In our discussion of pathways a few moments ago, we saw that they represent sequences of changes that occur in the world around us. To use an example from Atkins (1994, 184–185), we might have a biological reaction in which substrates are converted into product as follows:

[7] $$Rabbits + Grass \rightarrow more\ Rabbits,$$

and we might represent this in symbols as:

[8] $$R + G \rightarrow 2R$$

Take rabbits, add chemical energy photosynthesized from sunlight in the form of grass, and the result is more rabbits. The very presence of rabbits and an energy source to drive the process results in the production of more rabbits. Rabbits catalyze the production of more of themselves. The reaction type in pathway [8] represents what is known to students of chemical change as *autocatalysis*.

The rabbit reaction does not occur on its own, because the products feed into other reaction sequences:

[9] $$Rabbits + Foxes \rightarrow more\ Foxes,$$

or

[10] $$R + F \rightarrow 2F.$$

Foxes are hunted for their pelts to make coats for wealthy folk:

[11] $$Foxes \rightarrow Pelts \rightarrow Money,$$

or

[12] $$F \rightarrow P \rightarrow \$.$$

In this way, the furrier exploits a sequence of ecological pathways, powered by sunlight, to get a useful product. Just keep adding grass and let the system run. Chemists do the same thing in industrial processes driven by energy to convert substrates into saleable products. The populations that change for the chemist are populations made of molecules of various types. Autocatalytic chemical reactions are simply those whereby the very presence of a molecule of *X* catalyzes the formation of more *X*, thereby increasing the concentration of *X* in the reaction vessel.

Notice the way in which the reactions may be linked to form temporal structures in the form of oscillations—or cyclical changes over time. Rabbits beget more rabbits (through a well-known mechanism), and the rabbit population rises. The increasing rabbit population induces an increasing fox population, which consumes the rabbits, causing the rabbit population to collapse through overpredation. This change in turn reduces the fox population, and the process begins again, with rabbits proliferating in the absence of large numbers of predators. Cyclical changes like this, in interacting animal populations, have been seen many times by biologists, as have situations whereby steady states have been achieved. Sometimes seemingly chaotic changes can result, and the underlying order requires careful data analysis.

Suppose now that rabbits are introduced into a previously rabbit-free area. Autocatalysis results in lots of rabbits at the place of introduction. Since these are a nuisance, the Department of Agriculture might introduce foxes for biological control. As the initially localized population of rabbits becomes subject to predation, they migrate away, breeding all the while, to new locations. Assume that the rabbits move away in all directions from the place of introduction. The foxes follow behind, eating the products of all this reproduction, and the result will be an expanding wave of reproducing rabbits followed by an expanding wave of reproducing foxes. As the foxes deplete a given area of rabbits, they either move on or die, and the rabbit population in that region can recover, inducing more foxes to return.

The result, from an initial center of rabbits, will be concentric rings expanding outward—rabbits followed by foxes followed by rabbits. There will, in fact, be a changing spatial structure. Ecologists have found real systems similar to this, and we will later examine a chemical system (an example of what chemists call a *reaction-diffusion* system) that shows both spatial and temporal structure. Those of you who followed the discussion in the last chapter will no doubt have noticed how close we are to a discussion of evolution here, with all these references to predators, prey, survival, and reproduction. Lots and lots of reproduction.

To get closer to the issue of evolution, a thought experiment may help. Let's revisit the rabbit-grass pathway:

[13] $$R + G \rightarrow 2R$$

If the process was perfect and error-free, we might expect to see something like this:

[13a] $$R_{vo} + G \rightarrow R_{vo} + R_{vo}$$

The subscript *vo* stands for *variety of type-o*. Pathway [13a] indicates that *vo*-rabbits produce more rabbits of the same variety. Reproduction in accord with [13a] will result in a rabbit population consisting of individual rabbits of the *o*-variety. But rabbit reproduction is not a perfect process. Due to mutations, heritable changes creep in, and every now and again the process will have to be described differently as:

[13b] $$R_{vo} + G \rightarrow R_{vo} + R_{vi},$$

where the subscript *vi* indicates a new variant on the rabbit theme. Suppose now that the *vi*-rabbits can outrun *vo*-rabbits when chased by foxes (they don't have to outrun the foxes, only the *vo*-rabbits). Over many successive generations, as *vi*-rabbits increase in frequency while *vo*-rabbits decline in frequency due to predation, we will find that a better description of the rabbit-grass pathway (perhaps on average) is given by:

[13c] $$R_{vi} + \mathrm{G} \rightarrow R_{vi} + R_{vi}$$

Our population has evolved. The rabbit population, as a population of open-dissipative energy conduits, has shifted its salient characteristics.

The rabbit-fox pathway might have begun as:

[14a] $$R_{vo} + F_{vo} \rightarrow F_{vo} + F_{vo}$$

But as *vi*-rabbits come to predominate in the rabbit population, *vo*-foxes will find it harder to make a living. Every now and again, however, just by chance mutation, new variations on the fox theme will appear and we will get:

[14b] $$R + F_{vo} \rightarrow F_{vo} + F_{vk}$$

Suppose the *vk*-foxes are just a bit wilier or faster than *vo*-foxes. Then after many successive generations, we will perhaps best describe the process as:

[14c] $$R_{vi} + F_{vk} \rightarrow F_{vk} + F_{vk}$$

This brings out the important point that interacting populations coevolve.

The environment that organisms find themselves in has a non-living component (the abiotic environment) and a living component (the biotic environment). The living environment can be characterized in terms of predators, prey, pathogens, and parasites. Populations of organisms do not evolve independently of each other. Their evolutionary fates are typically coupled in interesting and complex ways. In fact, we have here what biologists refer to as *Red Queen coevolution*, named after the Red Queen in Lewis Carroll's *Alice through the Looking-Glass*, who had to run as fast as she could just to stay in the same place. The rabbit population changes, and the fox population changes to match it, and vice versa. The process doesn't end where we have left it in our very simplified example. It is ongoing and relentless.

Self-Organization and the Emergence of Order

Self-organized systems are complex, organized systems made up of many interacting subunits or parts. They are examples of open-dissipative systems. As energy and matter flows through such a system, the parts interact in such a way as to sustain the integrity of the system. In the process, the interactions among the parts give rise, collectively, to orderly, organized behaviors of the system as a whole. These system-level features are said to emerge from the interactions among the parts, and they are known as emergent properties. The coherent motions of the parts do not involve the intervention of an intelligence external to the system, nor do they need to arise from the operation of centralized control mechanisms internal to the system. The parts may simply be dumb molecules.

To get self-organization, several conditions need to be satisfied. These include the following:

(a) A Collection of Suitable Components

The components come in all sizes. They may be atomic or molecular (water will do), they may be cellular, they may be organismal, or they may even be the stellar components of galaxies self-organizing through gravitational energy into giant rotating spirals.

(b) Local Coupling Mechanisms

The components must be able to couple their behaviors (dynamics) in accord with local mechanisms.

It is this coupling of behaviors of components that lies at the heart of self-organization. Self-organized systems have many interacting parts whose interactions give rise to the global, collective behavior of the system. The dynamical process by which potential components of the system become integrated is the process by which the self-organized system emerges from the background. The resulting dynamical coupling gives rise to emergent global behaviors of the entire system (spatial and temporal patterns).

The requirement that components influence each other's behavior through causal mechanisms that act locally means that interactions among the parts must reflect causal mechanisms whose effects reflect purely local conditions as causes. The behavior of any component can only cause—and in turn be caused by—the behavior of its immediate neighbors. In this way, self-organizing systems do not require the existence of elaborate, systemwide communication systems—systems that would presuppose some degree of prior organization.

This local coupling of parts constrains their behavior, and their freedom to respond to changes in their immediate environments is thus restricted. This has the effect that the parts, thus constrained, can manifest coherent, nonrandom motions. This restriction also has the effect that a local environmental perturbation or disturbance in a self-organized system will tend to propagate through the system. The extent of the propagation will depend on the presence or absence of amplification mechanisms. (Autocatalysis, discussed previously, is an amplification mechanism that can be found in many self-organizing systems.) Damping mechanisms will also be important for the regulation of changes in self-organizing systems.

The stability of self-organizing systems results from the operation of regulatory mechanisms. Positive feedback (autocatalysis is an example) will tend to make a system grow through amplification of initial effects (as air is heated in a forming hurricane, it rises, drawing in more moist air, which surrenders its moisture, leading to the presence of more heat, which causes the air to rise even faster, which draws in even more moist air, etc.). But we do not see arbitrary growth, so positive feedback is balanced by negative feedback, which inhibits amplification. In the rabbit-grass pathway, we will not get an unlimited number of rabbits, because as they multiply, they consume grass. The availability of grass constrains the rabbit population growth (in ways relevant for evolution, as the resulting struggle for existence will favor some rabbit-variants at the expense of others).

A nice example of negative feedback in biochemistry concerns the pathway by means of which *Escherichia coli* bacteria synthesize the amino acid isoleucine from another amino acid, threonine (see Lehninger, Nelson, and Cox, 1993, 13). It is a five-step pathway:

$$[15] \qquad A \rightarrow B \rightarrow C \rightarrow D \rightarrow E \rightarrow F,$$

(Here, A = threonine, B = α-ketobutyrate, C = α-aceto-α-hydroxybutyrate, D = α, β-dihydroxy-β-methylvalerate, E = α-keto-β-methylvalerate, and $F =$ isoleucine). Each step in the pathway is catalyzed by a specific enzyme. Without regulation, so long as threonine was fed in, along with usable energy, isoleucine would be produced. But isoleucine levels do not rise arbitrarily, for the presence of increasing concentrations of isoleucine inhibits the first step $A \rightarrow B$. Isoleucine binds to the enzyme catalyzing this step, thus reducing its catalytic activity. In this way, rising levels of isoleucine regulate the rate of its own production, thus keeping cellular concentrations within acceptable limits.

(c) A Flow of Usable Energy

A flow of energy is needed to drive the formation of a self-organized system. This flow of energy into and out of the system must continue in order to sustain the system, driving the interactions among the components. A self-organized system starved of sustaining energy will sink back into the environment from which it emerged.

Self-organization thus occurs in systems taken out of thermodynamical equilibrium with their surroundings. Brian Goodwin, a developmental biologist who has studied the spatial and temporal structures and patterns resulting from self-organization in biological systems, has observed:

> What counts in the production of spatial and temporal patterns is not the nature of the molecules and other components involved, such as cells, but the way they interact with one another in time (their kinetics) and space (their relational order—how the state of one region depends on the state of neighboring regions). These two properties together define a field.... What exists in the field is a set of relationships among the components of the system. (1996, 51)

This field is sometimes referred to as an excitable medium because a collection of potentially interacting components may start out in a homogeneous state. (It may exhibit spatial and temporal symmetry, so one part looks pretty much like any other part.) This homogeneous condition will remain as the system is taken away from equilibrium by an input of usable energy. But the resulting nonequilibrium system is then poised to generate spatial and temporal patterns. It is said to be excitable. Excitation of the system, through the introduction of a local inhomogeneity, can break the initial spatial and temporal symmetry by inducing (through coupling of parts) excitations in adjacent parts of the medium, which in turn induce further excitations. By amplifying small fluctuations in the environment, positive feedback mechanisms can break the initial homogeneity of the excitable medium.

The result is that the initial disturbance propagates through the system, driving complex global behaviors of the system as a whole, because the behavior of any part of the system is constrained by the neighbors to which it is coupled (and their behavior in turn is similarly constrained). Recall, for example, that when the energetic conditions are right, a region of low pressure—an environmental inhomogeneity—can form the seed for the emergence of a hurricane. A hurricane, as we have seen, is a complex, self-organizing dynamical structure involving coherent motions of matter on an enormous scale.

The spatial and temporal order, patterns, and structure we can see in the behavior of self-organizing systems is not imposed from outside, nor does it arise from centralized control from within. The environment merely provides the energy to run the process, and environmental fluctuations are the usual sources of the initial local inhomogeneity that acts as a seed for the formation of the system in an initially homogeneous excitable medium. The patterns result from dynamical interactions internal to the system.

That there is evidently energy-driven interactive complexity in nature, giving rise to organized systems without intelligent design, there can be no doubt. And it is in this context that it is worth mentioning once again the distinction between appearance and reality that we discussed in the last chapter. As Seeley has recently noted, self-organization can give rise to the appearance of intelligence:

> We often find that biological systems function with mechanisms of decentralized control in which the numerous subunits of the system—the molecules of a cell, the cells of an organism, or the organisms of a group—adjust their activities by themselves on the basis of limited, local information. An apple tree, for example, "wisely" allocates its resources among woody growth, leaves, and fruits without a central manager. Likewise, an ant colony "intelligently" distributes its work force among such needs as brood rearing, colony defense, and nest construction without an omniscient overseer of its workers. (2002, 314)

Self-organization is not merely a process whereby complex organized systems can emerge and sustain themselves without intelligent design;

it is a process that can generate problem-solving systems out of dumb components, or out of components whose limited cognitive abilities are not up to the task of coordinating systemwide behaviors.

A good example here is afforded by the study of social insects. Colonies of social insects are open-dissipative systems. The component insects are dumb, yet by their mutual interactions they are capable of generating global, colony-level, problem-solving collective behaviors, with enormous implications for their survival and reproduction. The broader implications of these matters have recently been discussed under the heading of *swarm intelligence*. Thus Bonabeau, Dorigo, and Theraulaz observe:

> The discovery that SO (self-organization) may be at work in social insects not only has consequences on the study of social insects, but also provides us with powerful tools to transfer knowledge about social insects to the field of *intelligent system design*. In effect a social insect colony is undoubtedly a decentralized problem-solving system, comprised of many relatively simple interacting entities. The daily problems solved by the colony include finding food, building or extending a nest, efficiently dividing labor among individuals, efficiently feeding the brood, responding to external challenges, spreading alarm, etc. Many of these problems have counterparts in engineering and computer science. One of the most important features of social insects is that they can solve these problems in a very flexible and robust way: flexibility allows adaptation to changing environments, while robustness endows the colony with the ability to function even though some individuals may fail to perform their tasks. Finally, social insects have limited cognitive abilities: it is, therefore, simple to design agents, including robotic agents, that mimic their behavior at some level of description. (1999, 6–7)

Self-organizing systems made of unintelligent components can thus exhibit global, adaptive, purposive behaviors as a consequence of the effects of the collective interactions of their parts.

Moreover, these naturally occurring systems can serve as models that enable us to intelligently design artificial, soulless systems that will exhibit similar sorts of problem-solving activity. No ghost is needed in the collective machine, just interactions powered by usable energy in accord with mechanisms operating by the laws of nature. Prior to the study of self-organization, it used to be supposed either that social insects had some sort of collective "group-mind" that intelligently guided their behavior, or, alternatively, as Bonabeau,

Dorigo, and Theraulaz have noted, that individual insects possessed internal representations of nest structure, like human architects.

Neither assumption is warranted. The appearance of intelligent group behavior is the result of interaction dynamics internal to the colony of insects, duly modulated by environmental influences. Appearances can thus be deceiving. As Seeley has observed:

> No species of social insect has evolved anything like a colony-wide communication network that would enable information to flow rapidly and efficiently to and from a central manager. Moreover, no individual within a social insect colony is capable of processing huge amounts of information. (Contrary to popular belief, the queen of a colony is not an omniscient individual that issues orders; rather she is an oversized individual that lays eggs.) The biblical King Solomon was correct when he noted, in reference to ant colonies, there is "no guide, overseer or ruler" (Proverbs 6:7). (2002, 315)

We should not let our natural propensities for anthropomorphic thinking lead us into seeing intelligence and intelligent design where it does not exist.

Karsai and Penzes (1998, 2000), for example, have shown that the adaptive nest shapes of certain species of wasps emerge from simple rules governing the purely local interactions of individual wasps with each other and with the emerging nest structure. To build a compact nest, the wasps, unlike intelligent human architects, do not need to know the global shape of the nests, they do not need to measure the compactness of the structure, and they do not build the nest in such a way that the final shape is the end or goal of their behavior, either singly or collectively.

In other words, they do not build with a goal in mind. As a matter of fact, the emerging nest organizes its own construction as part of a self-organizing process in which the present state of the nest provides local cues to the dumb wasps about where to apply the next dollop of pulp. After the pulp is applied, this will change the local configuration of a given site on the nest, and this in turn changes the pattern of attractive local building positions on the developing nest. Karsai and Penzes have demonstrated that a wide variety of nest shapes, from complex twiglike structures to more spherical structures (depending on environmental circumstances), can be explained in this way.

Self-organization is not the only way to get complex structures. The simpler phenomenon of self-assembly is important, too. It is a process capable of producing organized three-dimensional structures, and its fruits may be of use to more sophisticated self-organizing systems. For example, proteins are made up of chains of amino acids. Which protein you get depends on the sequence of its component amino acids. But proteins achieve their biological functions, perhaps enhancing chemical reactions or inhibiting them, by virtue of their three-dimensional structure.

These three-dimensional structures result from elaborate and intricate folding. The folding is achieved through physical and chemical interactions between the amino acids in the sequence constitutive of the protein. Once the amino acids are present in sequence, the protein self-assembles its three-dimensional configuration. The folding does not require intervention by external mechanisms or agents. Systems of self-assembled proteins may then go on to interact among themselves either to form protein complexes or to self-assemble into more complex, nucleoprotein structures such as viruses (Gerhart and Kirschner 1997, 146). They may even participate in self-organizing biochemical systems. A good introduction to molecular self-assembly—in soap bubbles and proteins—can be found in Cairns-Smith (1986, 69–73). But there is an as yet unexplored source of order in biological systems that we must consider.

Ontogenetic Darwinism

The theory of evolution, which we discussed in the last chapter, is, in part, an explanation of the mechanisms that generate and preserve new varieties, thereby changing the structure of biological populations. It is also, in part, an account of the mechanisms by means of which new species come into being, and it is also our best explanation of the historical (phylogenetic) relationships exhibited by extant species. For want of a better term, I will call this strand of evolutionary thinking *phylogenetic Darwinism*. The crucial feature of phylogenetic Darwinism is that it operates in populations of organisms across generations. Darwin's *Origin of Species* is the first statement of phylogenetic Darwinism.

Though phylogenetic Darwinism came first from a historical standpoint, it has since become apparent that Darwinian principles operate within organisms in the course of their life cycles. Thus Gerhart and Kirschner have remarked:

> An alternative mechanism to self-assembly is to generate without strict bias a large number of possible states and select the most appropriate. Physiological systems based on variation and selection are much more prevalent in biology than has been appreciated. The power of Darwinian selection as a cellular mechanism in the short term (rather than a genetic selection mechanism used only in the long term) has recently become clearer. In many biological systems, several, and often a large number, of alternative responses to external stimuli are in fact produced, and one is selected. (1997, 147)

Since the study of ontogeny is the study of the development and life cycles of individual organisms, I will term these extensions of Darwinian ideas to events occurring within an individual in the course of its life cycle as *ontogenetic Darwinism.*

The immune system affords a good example of ontogenetic Darwinism in action in each of us. The example shows how Darwinism can be used to explain some important processes whereby individual organisms themselves become adapted to short-term changes in their environments (the sorts of changes that cannot be directly encoded and foreseen in the genome they inherit).

The immune response is the reaction of the body (self) to invasion by foreign substances (non-self) known as *antigens*. An important part of the immune response, known as *humoral immunity*, involves white blood cells known as *B* lymphocytes. These cells produce circulating proteins known as *antibodies*. Antibody molecules are referred to as *immunoglobulins* and are coded for by immunoglobulin genes. Antibodies react with antigens to flag them for further immunological action that (with luck) renders them harmless. (T lymphocytes are cells responsible for cell-mediated immunity. In this latter case, the immune response involves cells that are specially adapted to attack and destroy foreign bodies in an organism.) Cells of both types play a role in adaptive immunity.

I will focus here on *B* cells and the antibodies they produce and carry on their surfaces. The population of antibodies available to attack a given antigen will vary with respect to their ability to bind to that antigen. Some antibodies won't bind at all (or rarely), others will bind more frequently, and some will bind virtually every time they encounter the antigen. Antibodies are said to have specificity for the antigens to which they bind, and one of the things we will be concerned to discover is how this specificity is improved upon during the course of an infection. This will enable us to see how Darwinian mechanisms can tune an adaptive response within an individual.

To be effective, the immune system must produce an enormous range of antibodies. There are up to 10 billion *B* lymphocyte cells, and the system is capable of recognizing between 10 million and 100 million antigen shapes. What you inherit from your parents are immunoglobulin genes. As inherited, these are said to be in *germ-line configuration*. But what you inherit does not code for the immense diversity of antibody molecules. There is not enough information in the genome. In 1976, Susumu Tonegawa discovered that antibody genes are not inherited complete but rather as fragments that are shuffled together to form a complete immunoglobulin gene that specifies the structure of a given antibody. This process is known as *somatic recombination*, since it occurs in body cells that are not germ-line (reproductive) cells. As these fragments are combined to form a complete immunoglobulin gene, new DNA sequences are added at random to the ends of the fragments, ensuring still more antibody diversity.

This random reshuffling of immunoglobulin genes, together with the random insertion of DNA sequences during the somatic recombination process, results in a high probability that at least one antibody, though perhaps not binding perfectly, will fit at least one of the many determinants (molecular handles) presented by a new antigen. Once an antibody is selected by binding to antigen, it stimulates the *B* lymphocyte to which it is attached to make exact copies—clones—of itself. Some of these clones remain as circulating *B* lymphocytes, serving as the immune system's memory. Increased numbers of these cells provide a faster immune response to subsequent infections and establish the immunity that follows some infections and vaccinations. Other clones stop dividing, grow larger, and turn into plasma cells, whose sole function is to produce large numbers of free antibodies to fight the current infection. What about the observation that antibodies produced in the later stages of an infection are more effective at binding than the antibodies initially produced?

The fine-tuning of the antibody response is accomplished by another Darwinian mechanism that changes the genetic makeup of the immunoglobulin genes through mutation. This random mutation of the immunoglobulin genes is known as *somatic hypermutation*. By randomly producing many variations on a successful theme, some antibody variants will be better than the original clones at binding to a given antigen, and specificity will be enhanced (see Parham 2000, 21). This example shows dramatically how Darwinian mechanisms of mutation and selective retention of variants for further evolutionary modification results in the specificity of proteins that intelligent design theorists find so mysterious.

We have in the *B* lymphocyte population the random production of a wide range of variants, with differential reproduction—cloning—of selected variants, depending on the specific challenges to an individual's immune system. The clones of the selected variants inherit the genetic properties that made their progenitor cells successful. The immune system's adaptive response to novel antigen presentation is based on the same evolutionary principles that shaped the organism itself and adapted it to its external environment. Each of us has a unique immune system, whose current features reflect the historical contingency of fast evolution occurring during our life cycles.

In this way, our best theory of the immune system, with enormous implications for the way we think about infection, is thoroughly Darwinian. Thus, commenting on modern immunologists, Peter Parham observes:

> The very foundations of their subject are built upon stimulation, selection and adaptive change. Now we see clearly the immune system for what it is, a vast laboratory for high speed evolution. By recombination, mutation, insertion and deletion, gene fragments are packaged by lymphocytes, forming populations of receptors that compete to grab hold of antigen. Those that succeed get to reproduce and their progeny, if antibodies, submit to further rounds of mutation and selection. There is no going back and the destiny of each and every immune system is to become unique, the product of its encounters with antigen and the order in which they happen. This all happens in somatic tissues in a time frame of weeks. (1994, 373)

This is but one example. Others can be found in developmental biology, where the production of a superabundance of cells, with differential retention of a smaller number, plays a crucial role in developmental sculpting of such structures as fingers and toes. Other important examples can be found in neurobiology (Gerhart and Kirschner 1997) and in oncology (Greaves 2000).

In the nineteenth century, embryologists, noting analogies between the appearance of developing embryos (the human fetus goes through a gill-slit stage, for example) and major evolutionary events (fish evolved before amphibians, reptiles, and mammals), would sometimes use the slogan, *Ontogeny recapitulates phylogeny*. As a literal description of development, this slogan is horribly inaccurate. Yet at the very different level of the description of processes and mechanisms giving rise to complex, adaptive, problem-solving systems like the immune system, it may well be true that ontogeny does indeed recapitulate phylogeny. The variation-and-selection mechanisms driving adaptive evolution in populations of animals, for example, also operate at the level of populations of cells within our bodies in the course of our lifetimes.

With this background in thermodynamics and self-organization, we are now in a position to analyze the central claims of intelligent design theorists. I will begin here with a consideration of some incautious remarks about thermodynamics made by intelligent design theorist William Dembski.

Doubting Dembski: Misinformation and the Origin of Disorder

Intelligent design theorist William Dembski has tried to exploit thermodynamics in order to bolster his claims about intelligent design of nature. Indeed, he modestly claims to have discovered a fourth law of thermodynamics (2002, 169). What could this fourth law be, and how might it relate to the well-known Second Law? Dembski's candidate is something he calls the *Law of Conservation of Information*. To understand the proposed Law, we must see what Dembski means when he refers to something known as *complex specified information* (CSI for short). Dembski's central claim is that this sort of information is the hallmark of intelligent design in nature (see 2001b, 176).

Dembski tells us that to infer design in an object, pattern, or process we need to discern three things: *contingency*, *complexity*, and *specification*. In this context, "Contingency ensures the object in question is not the result of an automatic and therefore unintelligent process" (2001b, 178). Contingent objects, patterns, or processes must be consistent with the regularities (described by the laws of nature) involved in their production, but these regularities or laws must permit or be compatible with "any number of alternatives." Dembski explains that, "By being compatible with but not required by the regularities involved in its production, an object, event or structure becomes irreducible to any underlying physical necessity" (2001b, 178).

Complexity is said to derive from a low probability of occurrence of a pattern or process by chance alone. In particular, Dembski tells us, "Complexity and probability therefore vary inversely: the greater the complexity, the smaller the probability" (2001b, 179). Patterns of events that could easily happen by chance alone (for example getting two heads in two consecutive tosses of a fair coin) are not deemed to be complex. A pattern of events with a very low probability of happening by chance alone will, by contrast, be complex. I will call this kind of complexity *Dembski-complexity*.

In addition to contingency and complexity, we also need specification. Here we need to differentiate between specified patterns of events and purely *ad hoc* patterns. Dembski tells us, "For a pattern to count as a specification the important thing is not when it was identified, but whether, in a certain well-defined sense it is independent of the event it describes" (2001b, 182). CSI is the information contained in complex contingencies that "conform to an independently given pattern, and we must be able independently to construct that pattern" (2001b, 189).

Thus, to use a variant of one of Dembski's own examples, a gunman who shoots at a wall and who then draws bull's-eyes around the bullet holes will have generated an *ad hoc* pattern, and not a specified pattern. The pattern of hits is not specified in advance (or independently) of the shooting that gave rise to them. One who shoots once at a fixed target and hits the bull's-eye may simply be lucky. By contrast, it is Dembski's idea that a gunman who shoots many bullets from a distance at a fixed target and who hits the bull's-eye each time will have generated a contingent, complex, and specified pattern of events. Such a pattern is not the result of an automatic process, it has low probability of occurring by chance alone, and it conforms to an independently specified pattern. Dembski claims that these types of patterns contain the trademarks of intelligent design—in the present case, that the pattern has arisen from the skill of the gunman who has intelligently designed the trajectories of his bullets.

I have argued in this chapter that self-organization can give systems the *appearance* of being intelligently designed while being in reality the result of dumb, natural mechanisms. It is a good question as to whether the fruits of undesigned self-organizing processes pass muster as being intelligently designed when viewed from the standpoint of Dembski's claims about CSI. If they do, then the design inference will be invalid: for an enormous number of natural systems it will lead from *true* premises concerning contingency, complexity, and specification to *false* conclusions about intelligent design. An example may help.

A natural phenomenon involving self-organization that can easily be reproduced in the laboratory concerns Bénard convection cells (figure 3-2). Consider a thin layer of water sandwiched between two horizontal glass plates. Suppose the system is at room temperature and in thermal equilibrium with its surroundings. One region of water looks pretty much the same as any other. If the water is now warmed from below, so that energy is allowed to flow through the system, and back into the environment above, there is a critical temperature where the system will become self-organized. In this case, this means that if you look down at the system, you will see a structured, honeycomb pattern in the water.

The cells in the honeycomb—often appearing as hexagons or pentagons—are known as Bénard cells, and are rotating convection cells. Water warmed from below rises; as it rises heat dissipates, and the water cools and starts to sink again to the bottom to be re-warmed, thereby repeating the process (figure 3-3). Water cannot both rise and fall in the same place, so regions where water rises become differentiated from regions where it sinks. It is this differentiation that gives rise to the cells. The cells have a dimpled appearance, since water rises up the "walls" of the cell and flows toward the center "dimple" to flow back down again, completing the convective circulation.

The cells are visible because of the effects of temperature on the refraction of light. The way one cell rotates influences—and in turn is influenced by—the ways in which its immediate neighbors rotate.

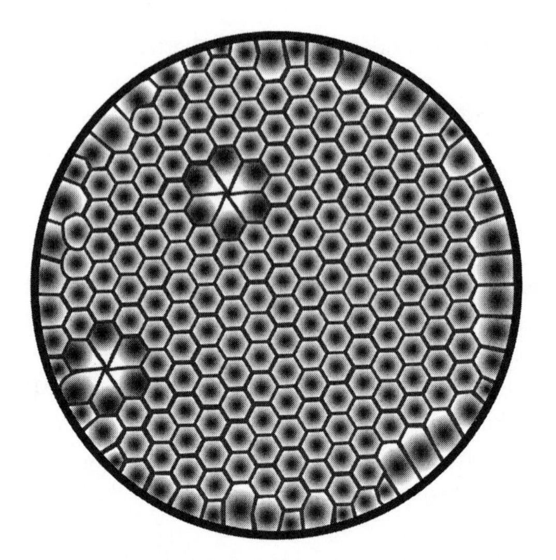

Figure 3-2. Bénard convection cells in a Petri dish. The cells have similar size (except on the border) and although the shapes of the cells vary they contain hexagonal structures similar to those found in a honeycomb. The emergence of an organized structure from a homogenous medium like water or oil is quite startling.

By adding thermal energy to water, we have brought about the spontaneous emergence of a complex system of mutually interactive convection cells. And not just in water, for astronomers have seen these cells on the surface of the sun, another well-known system far from equilibrium.

The spatial and temporal organization and patterns we can see in the behavior of this self-organizing system is not imposed from outside by a designing intelligence. The environment merely provides the energy to run the process. Environmental fluctuations are the usual sources of the initial local inhomogeneity that acts as a seed for the emergence of the system in an initially (almost) homogeneous aqueous medium. The patterns result from the energy-driven interactions of the components internal to the system.

Apparently aware of the threat posed by self-organization of this kind for his attempted defense of claims about intelligent design, Dembski initially accuses those who study these phenomena of trying to get a free lunch,

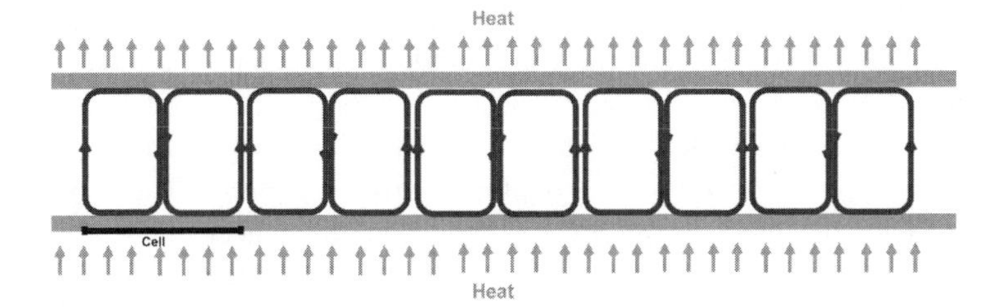

Figure 3-3. Bénard convection cells in cross section. Consider the cell above the horizontal black bar. The outer "wall" of the cell is generated by warm water rising. The water cools as it rises, and so flows to the center "dimple," where it sinks back down to be warmed again, thereby repeating the process.

> Bargains are all fine and good, and if you can get something for nothing, go for it. But there is an alternative tendency in science that says that you get what you pay for and that at the end of the day there has to be an accounting of the books. Some areas of science are open to bargain-hunting and some are not. Self-organizing complex systems, for instance, are a great place for scientific bargain-hunters to shop. Bénard cell convection, Belousov-Zhabotinsky reactions, and a host of other self-organizing systems offer complex organized structures apparently for free. But there are other areas of science that frown on bargain-hunting. The conservation laws of physics, for instance, allow no bargains. (2001a, 23)

Dembski does not tell us which conservation laws of physics forbid self-organization. This is a vexing matter since Bénard cells occur in nature—for example in the sun—as well as in the laboratory (not to mention a host of other self-organizing systems). *Their existence is certainly consistent with known conservation laws.* Not only this. For Bénard cells, forming in accord with dumb, natural mechanisms, manifest *complex specified information.*

First of all, Bénard cells manifest Dembski-complexity. The formation of Bénard cells just by chance alone is highly improbable. In fact they do not form *just by chance*. The cells result from self-organizing processes whose physical consequences are the emergence of visible patterns involving the net coherent, coordinated motions of trillions of water molecules. (Just eighteen grams of water, one mole, contains 6.02×10^{23} water molecules.) The patterns are thus extremely complex. The general pattern can also be specified independently of, and indeed prior to, its generation. The patterns are thus not *ad hoc*. Bénard cells are also contingent. They do not result from an automatic process that gives you the exact same pattern each time.

The situation with respect to both contingency and specification is similar to the one we just encountered with our gunman. The gunman intelligently designs the trajectories of his bullets to hit the bull's-eye from a great distance. Being a skillful gunman, he hits the bull's-eye every time. The general specified pattern is a pattern of hits in the region of the bull's-eye that has a low probability of happening by chance alone. But each time the gunman shoots a sequence of several bullets in order to demonstrate his skill he gets a different pattern of hits in the region of the bull's-eye of the target. (If he got exactly the same pattern every time this would call contingency into question, and we might suspect an automatic, as opposed to a skillful, process was at work.)

Like the marksman's pattern of hits on the target, the general Bénard cell pattern involves *some* arrangement of interacting pentagons and hexagons each time you run the experiment. But you never get exactly the *same* pattern (i.e., arrangement of hexagons and pentagons, along with their mutual rotational interactions) twice—it is nothing like an automatic process that gives the exact same result, *repeatedly and reliably*, each time. In this regard, Dembski (2002, p. 243) is guilty of gross oversimplification in his desire to quickly dispose of the problem posed by Bénard cell patterns. The crucial difference between the Bénard cell pattern and the pattern of hits by the marksman is that the Bénard cell pattern does not require intelligent design for its appearance, *only a dumb generating mechanism combined with the effects of dumb chance in the form of fluctuations and inhomogeneities in the dumb aqueous medium.*

What then of this new conservation law Dembski has told us he has discovered? According to Dembski, the *Law of Conservation of Information* is captured by the claim that *natural causes cannot generate CSI*. He lays out its implications as follows:

> Among its immediate corollaries are the following: (1) The CSI in a closed system of natural causes remains constant or decreases. (2) CSI cannot be generated spontaneously, originate endogenously or organize itself (as these terms are used in origins of life research). (3) The CSI in a closed system of natural causes either has been in the system eternally or was at some point added exogenously (implying that the system,

> though now closed, was not always closed). (4) In particular any closed system of natural causes that is also of finite duration received whatever CSI it contains before it became a closed system. (1999, 170)

We will see below that all these claims are false, and since they are alleged to be corollaries of the proposed "law" of conservation of information, we must conclude that it too is false.

First of all, Bénard cells manifest CSI and they arise from natural unintelligent causes. Moreover the central issue is not whether the system manifesting CSI is a component of a closed thermodynamical system. The central issue is whether there is usable energy to drive the formation of systems manifesting CSI. The universe we live in clearly does contain such usable energy, and is in fact teeming with such undesigned yet organized complex systems at all scales, from the molecular to the galactic.

In view of the fact that self-organization can give rise to systems manifesting CSI we must now reexamine Dembski's theoretical account of the relation of his proposed law of conservation of information to the well-established laws of thermodynamics. To this end, Dembski notes of his proposed law:

> Moreover, it tells us that when CSI is given over to natural causes it either remains unchanged (in which case the information is conserved) or disintegrates (in which case information diminishes). For instance, the best that can happen to a book on a library shelf is that it remains as it was when originally published and thus preserves the CSI inherent in the text. Over time, however, what usually happens is that a book gets old, pages fall apart, and the information on the pages disintegrates. *The Law of Conservation of Information is therefore more like a thermodynamic law governing entropy, with the focus on degradation rather than conservation.* (2002, 161–162, my italics)

But exactly how is this proposed law *like* a thermodynamical law-governing entropy? What sort of relationship is being claimed here?

In an attempt to clarify the relationship between his proposed law and the accepted laws of thermodynamics, Dembski wonders,

> . . . whether information appropriately conceived can be regarded as *inverse to entropy* and whether a law governing information might correspondingly *parallel the second law of thermodynamics, which governs entropy*. Given the previous exposition it will come as no shock that

> my answer to both questions is yes, with the appropriate form of information being complex specified information and the parallel law being the Law of Conservation of Information. (2002, 166–167, my italics)

In saying that information can be thought of as being inverse to entropy, Dembski is arguing that as the entropy of a system decreases, information increases, and as entropy increases, information decreases.

To this last claim about the relationship between thermodynamics and entropy Dembski adds the following qualification:

> CSI, whose source is ultimately in intelligence, can override the second law. It is not fair to call this overriding of the second law a violation of it. The second law is often stated nonstatistically as the claim that in a closed system operating by natural causes entropy is guaranteed to remain the same or increase. But the second law is properly a statistical law stating that in a closed system operating by natural causes, entropy is overwhelmingly likely to remain the same or increase. The fourth law, as I am defining it, accounts for the highly unlikely exceptions. (2002, 173)

This passage gets us to the heart of the matter. To see why, we must have a brief excursion into the history of cosmology. (A fuller discussion of these matters will be undertaken in chapter 6.)

The scientists who developed the basic ideas of thermodynamics in the nineteenth-century tried to tease out the implications of this branch of science for the nature of the universe. The nineteenth-century physicist Ludwig Boltzmann was the first scientist to argue that the Second Law of Thermodynamics was a statistical law. In these terms, the entropy of an isolated or closed system will tend increase until it attains a state of thermodynamical equilibrium, at which point the entropy will tend to remain unchanged. Exceptions to these trends with respect to entropy were claimed to be due to the occurrence of random fluctuations bringing about spontaneous decreases in entropy, with large fluctuations being much more unlikely than very small fluctuations.

Boltzmann was driven to the view that the ordered, structured universe we see around us was due to an enormous, incredibly rare statistical fluctuation that had brought about a massive, spontaneous decrease in entropy. Even in its own terms, this explanation of the organized character of the universe we live in is not satisfactory. As astrophysicist Martin Rees has observed: "Indeed Boltzmann should have concluded that his brain was receiving coordinated stimuli that gave the illusion of a coherent external world which didn't actually exist. This solipsistic perspective would be vastly less improbable than the emergence of the whole external world as a random fluctuation!" (1997, 221) Put this way one might be tempted to forgive Dembski for his claims that the complex, structured universe we see results from the intelligent designs of a being outside the system. Such charity would be premature. The real problem is that neither Dembski nor Boltzmann are correct about the nature of the universe we live in. This has the consequence that Dembski's proposed fourth law of thermodynamics, the law of conservation of information, is simply not needed to *explain* the "highly unlikely exceptions" to the Second Law that Boltzmann had attributed to random fluctuations!

What sort of a world do we live in from a thermodynamical point of view? Rees has observed:

> The everyday world is very far from thermal equilibrium—there are enormous contrasts between hot and cold. It is not completely ordered; nor has it "run down" to a completely disordered and random state. The same is true for the cosmos on larger scales—there are huge contrasts between the stars with their blazing surfaces (and still hotter centers) and the sky between them, which is almost at the absolute zero of temperature—not quite, of course, because it is warmed to 2.7 degrees by the microwave "echoes" from the big bang. In the ultimate future . . . conditions may revert closer to equilibrium, but this will take immensely long even compared with the universe's present age. (1997, 212)

As this passage makes clear, our universe is *presently* a nonequilibrium universe in which there is plenty of usable energy to drive the formation of organized structures on both small and large scales. But there is more. Our universe began with a Big Bang. Exactly what this means will be discussed in chapter six, but the following remarks are relevant here.

In the Big Bang cosmology of modern science, the entire universe (matter, energy, space and time) was originally scrunched up into a featureless, pointlike object known as a *singularity* that lacked structure and organization. In the beginning, entropy was *not* at a minimum, instead *entropy was at a maximum*. A universe with features, structure, and organization had to evolve out of this initial, maximally entropic condition. How could our universe have evolved away from equilibrium into a more structured, feature-filled, organized condition? Doesn't this very suggestion violate the Second Law and the "requirement" that everything should be running downhill from an initially ordered state? If the universe had a fixed volume, we might have a problem. But it does not, and so we must re-examine these entropy issues from the standpoint of the effects of gravity in an expanding universe. In essence, the expansion of the universe from an initial point-like singularity creates opportunities for gravity to initiate self-organizing processes the structured, feature-filled fruits of which are themselves the basis for further self-organization and emergence of additional structure and order.

In an expanding universe like ours which began smaller than an atom but with a Big Bang, gravity amplifies tiny (quantum) inhomogeneities in the density of the expanding universe to allow stars (and solar systems) and galaxies to form from an *almost* homogeneous background. Explaining this idea, Rees has noted of the early universe that,

> Any patch that starts off slightly denser than average, or is expanding slightly slower than average, would decelerate more because it feels extra gravity. Its expansion lags further and further behind, until it eventually stops expanding and separates out as a gravitationally bound system. This process is, we believe, what allowed galaxies and stars to form about a billion years after the Big Bang. (2001, 76)

In this context, self-organization plays a crucial role. It accounts simultaneously for the emergence of structure, pattern, and organization on both a cosmic and a local scale. How?

Slightly denser regions of interstellar gas (hydrogen and helium were the principal fruits of the Big Bang) gravitationally attract gas from their surroundings, thereby increasing their density and their ability to attract more gas in this way. As more gas is drawn in, a dense, rotating ball of gas forms that starts to heat itself by gravitational compression. In essence its own gravity makes it fall in on itself and it heats up as a result. In this process, a point is reached where the temperatures and pressures at the core of the gas ball are such as to result in the initiation of nuclear fusion reactions (similar to those that occur in a hydrogen bomb). At this point the gas ball has self-ignited to become a star. Fusion reactions in the lifetimes of stars (including those in their sometimes fiery deaths as supernovae) take lighter elements and fuse them into heavier elements, ranging from gases such oxygen and nitrogen, to heavy metals such as lead, gold, and uranium.

At large scales, gravity unites stars into galaxies. At small scales, nuclear reactions in the hearts of stars make the heavier elements in the periodic table. Astrophysicist Craig Hogan has observed:

> These elements contributed most of the solid particles that accumulated into rocky planets like ours. In the formation of a star, rotation forces the gas into discs like miniature galaxies, which eventually become planetary systems as the material in them collects into planets. Because of the heat close to the main star, all that is left is the stuff that is heavy and hard to boil away; this is why the Earth has almost no helium and has hydrogen only in molecular combination with heavier atoms. More distant and massive "gas giant" planets, such as Jupiter, Saturn, Uranus, and Neptune, have hydrogen- and helium-rich composition like that of the sun. (1998, 128)

Gravity makes gas and dust form rocky planetesimals, and these, again under the influence of gravity, form planets. At least one of these planets, warmed by a large hot star, has been both a source of raw materials and a location for the self-organization of complex molecules and subsequent organizations of these in turn into the more complex structures we know of as *evolving life*. While much is still uncertain about the origins of life on Earth, we are now beginning to understand how the building blocks of complex biological molecules may have formed, and how these in turn may have organized into more complex structures. A good review of our current understanding of these matters can be found in Lahav (1999).

Our universe began in a state of maximal entropy. Subsequent self-organization has resulted in the formation of localized islands of order, structure, and complexity. The universe we live in has lots of usable energy and is far from a state of thermodynamical equilibrium in which entropy and information would remain the same (barring statistical fluctuations). The resulting *decreases in entropy* in these islands of order, by Dembski's own admission that entropy and information are inversely related, result *in increases in information*. These features of our universe point clearly to the conclusion that *you can indeed get CSI through self-organization resulting from unintelligent natural causes, and that no invisible supernatural hand operating outside a system of purely natural causes is needed.*

Self-organization is indeed a great scientific bargain when compared with evidentially empty promissory notes concerning supernatural design from outside our natural universe. (The reader seeking reviews treating other aspects of Dembski's musings on the topic of intelligent design should consult Elsberry 1999 or Shallit 2002.)

While natural modesty prevents me from proposing my own fourth law of thermodynamics, I would like to suggest that those with a taste for naming laws of nature at least have the common decency to ensure that they do not add appreciably to entropy (measured in incoherent disorder) of the intellectual universe in which we dwell as open-dissipative thinking systems struggling desperately to make sense of the world around us. Dembski's musings about thermodynamics and intelligent design amount to little more than putting information-theoretic lipstick on an old creationist pig.

In this chapter we have seen that there are many routes to complex, organized, structured physical systems. The Second Law, far from being an obstacle to evolution, provides us with a deep understanding of it. But now we must examine more serious allegations by intelligent design theorists. First they have alleged that science by its very nature is prejudiced against appeals to supernatural beings and supernatural causation. Second they have alleged that there are biochemical adaptations in organisms that defy explanation in natural terms and require supernatural intelligent design for their explanation. Third they have alleged that modern cosmology reveals a universe that requires for its existence a supernatural intelligent designer. These matters will be addressed in the next three chapters.

4

Science and the Supernatural

So far we have examined the development of intelligent design theories, we have examined the development of the theory of evolution, and we have shown how self-organization, occurring in accord with the laws of thermodynamics, can give rise to ordered complex structures in living and nonliving systems without a need for supernatural intervention. The time has now come to examine the modern intelligent design movement in detail. This will involve an examination of claims that important biochemical systems could not have evolved and must be the fruits of intelligent design. These claims will be examined in the next chapter. We must also examine the claim that the universe itself shows evidence of intelligent design. This will be done after we have examined intelligent design in biochemistry. But before turning to these matters, we must examine the central claims of contemporary intelligent design theory. What is it? What does it seek to accomplish? How does it differ from natural science? These questions are the main business of this chapter.

As we saw in the introduction, intelligent design theorists are pursuing a wedge strategy. The architect of this strategy is Phillip Johnson, who has observed:

> Our strategy is to drive the thin end of our Wedge into the cracks in the log of naturalism by bringing long-neglected questions to the

> surface and introducing them into public debate. Of course the initial penetration is not the whole story, because the Wedge can only split the log if it thickens as it penetrates. . . . A new body of research and scholarship will gradually emerge, and in time the adherents of the old dogma will be left behind, unable to comprehend the questions that have become too important to ignore. (2000a, 14–15)

At the thin end of the wedge, we can find opposition to naturalism. As the wedge thickens, there is a spirited defense of the argument from design. As we will see here and in the conclusion to this book, there are some very disturbing claims at the fat end of the wedge. We will begin with issues at the thin end of the wedge.

The Critique of Naturalism

The new intelligent design movement claims that science has been taken over by a pernicious, atheistic philosophy whose names are legion: naturalism, materialism, physicalism, and modernism. Phillip Johnson puts it this way:

> Under any of those names this philosophy assumes that in the beginning were the fundamental particles that compose matter, energy and the impersonal laws of physics. To put it negatively, there was no personal God who created the cosmos and governs it as an act of free will. If God exists at all, he acts only through inviolable laws of nature and adds nothing to them. In consequence, all the creating had to be done by the laws and particles, which is to say by some combination of random chance and lawlike regularity. It is by building on that philosophical assumption that modernist scientists conclude that all plants and animals are the products of an undirected and purposeless evolutionary process—and that humankind is just another animal species, not created uniquely in the image of God. (2000a, 13–14)

On this view, philosophy and not evidence is what underlies scientific support for evolution. While I think the claim that the theory of evolution rests on pernicious philosophy is false, it is at least something worth arguing about.

Johnson (2001, 29) himself sees the problem as lying in what he sees as the schizophrenic character of modern science. He alleges that there are two strands to modern science—two models, if you will—undergirding the practice of science. Johnson sees these strands as being intertwined, and they need to be separated carefully. Both strands have roles in the debate about Darwinism and intelligent design. The first strand or model concerns materialism or naturalism, which Johnson thinks is a philosophical theory that scientists have assumed without good reason:

> Within this first model, to postulate a non-material cause—such as an *unevolved intelligence* or *vital force*—for any event is to depart altogether from science and enter the territory of *religion*. For scientific materialists, this is equivalent to departing from objective reality into subjective belief. What we call intelligent design in biology is by this definition inherently antithetical to science, and so there cannot conceivably be evidence for it. (2001, 29, my italics)

I note here that it is interesting that Johnson evidently considers intelligent design to be on a par with the old idea of vitalism and its references to vital forces. This is something that deserves further scrutiny.

Vitalism was an idea with ancient roots that was prevalent, like intelligent design, in the eighteenth and early nineteenth centuries. It is perhaps better described as a collection of theories put forward to explain the differences between living and nonliving systems. Living systems were said to be animated by vital forces, the so-called élan vital. In the nineteenth century, as modern chemistry started to mature, vitalism evolved into the idea that the organic chemicals constitutive of organisms could be made only inside organisms, because only organisms possessed the vital force needed for organic synthesis. The idea was dealt an early blow by Friedrich Woehler, who, in 1828, synthesized an organic substance (urea) in the laboratory, without the aid of an organism and its alleged vital forces.

The science of thermodynamics, which emerged in the nineteenth century, was also relevant. Since ancient times, people had wondered about the heat generated by animals (including ourselves). How did live, warm-blooded animals produce heat? How did they stay warm instead of cooling gradually as rocks do once they have been heated? Dead animals were cold. Yet no obvious combustion was evident inside living creatures—no fires burned and smoked in bellies. Vitalists thought the heat was a by-product of the operation of vital forces that accounted for the difference between living and nonliving systems. But this idea was dealt a series of scientific blows.

First, the chemistry of oxidation was gradually unraveled, enabling eighteenth-century chemists like Lavoisier to understand the chemical basis of respiration (oxygen breathed in, carbon dioxide breathed out). Second, as the Law of Conservation emerged in the nineteenth century, it became apparent that energy, while it could be neither created nor destroyed, could change its form. Writing around 1852, Robert Mayer could observe: "Carbon and hydrogen are oxidized and heat and motive power produced. Applied directly to physiology, the mechanical equivalent of heat proves that the oxidative process is the physical condition of the organism's capacity to perform mechanical work and provides as well the numerical relations between [energy] consumption and [physiological] performance" (quoted in Coleman 1977, 123). The chemical energy in food could be converted through chemical action into the mechanical energy and heat energy observed in animals. There was a combustion of sorts after all but one that could be understood in natural, chemical terms without references to mysterious vital forces.

Scientists eventually lost interest in vitalism because there was no evidence to support its central claims (no vital forces were ever measured) and because the very phenomena that seemed to call for vitalism could be given good scientific explanations without reference to vital forces. Despite current attempts to revive intelligent design, we will see that it, too, has similar evidential and explanatory defects.

Notwithstanding this, for Johnson evidence is the central issue. Accordingly, he notes that the second strand or model underlying scientific practice is the empirical model that does not exclude nonphysical entities—for example supernatural entities—from the outset of inquiries but requires that hypotheses be formulated and tested and that data be fairly examined. Johnson observes:

> Within science one cannot argue for supernatural creation (or anything else) on the basis of ancient traditions or mystical experiences, but one can present evidence that unintelligent material causes were not adequate to do the work of biological creation. Whether some phenomenon could have been produced by unintelligent causes, or whether an intelligent cause must be postulated, both ideas are eligible for investigation whether the phenomenon in question is a possible prehistoric artifact, a radio signal from space, or a biological cell. (2001, 29)

He adds:

> Scientific empiricists, as I use the term, hold there are three kinds of causes to be considered rather than only two. Besides chance and law, there is also agency, which implies intelligence. Intelligence is not an occult entity, but a familiar aspect of everyday life and scientific practice. No one denies that such common technological artifacts as computers and automobiles are the product of intelligence, nor does anyone claim that this fact removes them from the territory of science and into that of religion. (2001, 30)

The emphasis on the importance of evidence is laudable, and we shall examine much that is offered to support the claims of intelligent design hypotheses in the rest of this book.

Though Johnson takes pains to give the appearance that he is taking the evidential high road, what Johnson is careful *not* to discuss is the possibility that the real reason scientists reject the hypotheses of intelligent design theory, like the vitalistic hypotheses of nineteenth-century biology, is precisely because the claims of intelligent design theory, like those of vitalistic biology, have absolutely no evidential support. Pointing to the existence of intelligent design by humans in the context of automobile or computer manufacture is utterly irrelevant to the issue of whether there was supernatural design of life or the universe itself.

The central stumbling blocks for intelligent design theory actually have little to do with pernicious materialistic philosophies alleged to be held by its opponents. The central stumbling blocks are all evidential in nature. The accusation that scientists reject intelligent design theory because they are in the sway of materialistic or naturalistic philosophy is part of a smoke-and-mirrors strategy to cover this sad reality from public scrutiny. For this reason, we must examine these matters more closely.

The Nature of Naturalism

To the claim that modern science rests on a pernicious naturalistic philosophy, scientists have objected that while some individual scientists may have a naturalistic philosophy, science as an activity has no such commitment. Instead, they say, science itself is committed to something quite distinct, called *methodological naturalism*. Intelligent design theorists have indeed been beaten with some considerable justification with this stick. In unpacking the distinction between philosophical naturalism, on the one hand, and methodological naturalism on the other, I will begin with the way intelligent design theorists see these matters, and then I will try to work back to something more reflective of reality.

For intelligent design theorist William Dembski, naturalism has had an odious influence in religion as well as science: "Hindu pantheism is perhaps the most developed expression of religious naturalism. In our Western society we are much more accustomed to dealing with what is called scientific naturalism. Ironically scientific materialism is just as religious as the overt religious naturalism of Hinduism. . . . Naturalism leads irresistibly to idolatry" (1999, 101). Philosophical naturalism, be it religious or scientific, has serious theological consequences:

> For those who cannot discern God's action in the world, the world is a self-contained, self-sufficient, self-explanatory, self-ordering system. Consequently they view themselves as autonomous and the world as independent of God. This severing of the world from God is the essence of idolatry and is in the end always what keeps us from knowing God. Severing the world from God, or alternatively viewing the world as nature, is the essence of humanity's fall. (1999, 99)

At this point, we seem to have left science behind in favor of theology and mysticism. This matter is important because nowhere does Dembski offer the slightest shred of evidence for his claims about humanity's fall, and with good reason: There isn't any. There is only religious faith.

What then is methodological naturalism, and what is its relation to metaphysical naturalism? Dembski describes methodological naturalism and its implications for intelligent design theory as follows: "The view that science must be restricted solely to undirected natural processes also has a name. It is called *methodological naturalism*. So long as methodological naturalism sets the ground rules for how the game of science is to be played, intelligent design has no chance of success" (1999, 119). Methodological naturalism so characterized emerges as a conceptual bogeyman out to thwart the honest endeavors of intelligent design's godly advocates. Dembski continues:

> We need to realize that methodological naturalism is the functional equivalent of a full-blown metaphysical naturalism. Metaphysical naturalism asserts that nature is self-sufficient. *Methodological naturalism asks us for the sake of science to pretend that nature is self-sufficient.* But once science is taken as the only universally valid form of knowledge within a culture, it follows that methodological and metaphysical naturalism become functionally equivalent. What needs to be done, therefore, is to break the grip of naturalism in both guises, methodological and metaphysical. (1999, 119–120, my italics)

But this characterization of methodological naturalism is a straw man—a position not actually maintained by theorists committed to methodological naturalism. It is a phantom in the minds of the advocates of intelligent design. First, methodological naturalism does not ask us to pretend that nature is self-sufficient. Second, methodological naturalism is not functionally equivalent to metaphysical naturalism.

Contrary to Dembski's gross and egregious mischaracterization of methodological naturalism, it is in fact a position that respects the gathering of good scientific evidence and the consequences of such evidence for our thinking, once gathered. Methodological naturalism, as it appears in science, is based on an inductive generalization derived from 300 to 400 years of scientific experience. Time and time again, scientists have considered hypotheses about occult entities ranging from souls, to spirits, to occult magical powers, to astrological influences, to psychic powers, ESP, and so on. Time and time again such hypotheses have been rejected, not because of philosophical bias, but because when examined carefully there was not a shred of good evidence to support them. Scientists are allowed, like anyone else, to learn from experience. Hard-won experience in the school of empirical hard knocks leads to methodological naturalism. The experience is straightforward. We keep smacking into nature, whereas the denizens of the supernatural and paranormal realms somehow manage to elude careful analysis of data.

Thus an important functional difference between methodological naturalism and metaphysical naturalism is this: The methodological naturalist will not simply rule hypotheses about supernatural causes out of court, as would a metaphysical or philosophical naturalist. But the methodological naturalist will insist on examining the evidence presented to support the existence of supernatural causes carefully and will ask—as is part of standard scientific practice—whether there are alternative explanations that will explain the same phenomena, especially less exotic explanations grounded in natural causes—the sorts of causes we have good reason to accept because we have bumped into them and their consequences time and time again in science and everyday life. The methodological naturalist will also be concerned with an examination of methods used to gather data. Were the methods used up to the task in question? and so on. This is part and parcel of everyday scientific activity.

With this in mind and by virtue of long scientific experience in which hypotheses about the supernatural, the magical, and the occult have failed to hold water, the methodological naturalist will view such hypotheses in the future with extreme caution (the same sort of caution we apply to alchemists who claim to be able to turn base metals into gold and to Realtors who claim to have a bridge in Brooklyn for sale at a reasonable price). Our caution simply reflects the experiences we have learned from. But methodological naturalists do not rule out the supernatural absolutely. They have critical minds, not closed minds. That we can simply rule out claims about the supernatural without further consideration is what metaphysical or philosophical naturalism is all about. Metaphysical naturalism is simply not the same as methodological naturalism.

This advice about the importance of critical thinking is not simply stored in the closet of science just for hypotheses about the supernatural or paranormal. The advice forms part and parcel of what good science is all about. What is known as *junk science* can just as well be science about natural rather than supernatural phenomena that fails the test of critical scrutiny when examined carefully for evidential and methodological defects. Cold fusion, with its promise of cheap, easy-to-obtain power, gained much public attention back in the late 1980s, yet it has since largely faded from public view precisely because the evidence presented, and the methods used, were not up to the task of demonstrating what was claimed by the central figures behind the idea. The scientific community is also interested in scientific fraud, whereby results are generated dishonestly with a view to deceiving the scientific community and the public. Forensic investigations into fraud use the same critical standards that are characteristic of methodological naturalism.

The issues here do not center on just junk science, scientific fraud, and hypotheses about the supernatural. The critical standards employed in these investigations are the same standards that have routinely led to the downfall of highly cherished ideas in science—ideas about natural things that were strangely elusive. Thus, back in the late eighteenth century, chemists stopped talking about phlogiston, the hypothetical substance of fire, to explain combustion. They stopped seeing combustion as a process whereby phlogiston was emitted from a combustible substance when it burned and came, instead, to see combustion as a process whereby oxygen unites with the combustible substance—a process known as *oxidation*, with the product of combustion being known as the *oxide*.

This change was not driven by philosophical antiphlogiston prejudice by the scientific elites. It was driven by data. The oxide weighs more than the original substance prior to oxidation. That the oxide was heavier was explained by phlogiston's advocates as a consequence of phlogiston having negative weight (so things got heavier as it was emitted). Lavoisier, the Isaac Newton of chemistry, realized that if phlogiston was made of matter, as its proponents argued, it had mass and hence, thanks to gravity, it necessarily had positive weight. If phlogiston was emitted, the resulting oxide, contrary to experience, would have to be lighter. The explanation lay elsewhere in oxygen, which also had mass and hence weight. This is why the oxide weighs more than the substance that was oxidized. This hypothesis was vindicated by careful experiments.

Similar stories can be told about the fall of such cherished ideas in science as caloric (the substance of heat), vital forces, and luminiferous aether. This last case is of particular interest in the present context. To understand what is involved here, consider ripples in water spreading out in a pond. The ripples are waves that propagate or "travel" across the pond. Water is said to be the medium of propagation for these waves. By the end of the nineteenth century, many physicists were convinced that light was composed of waves that traveled through space. But waves of any kind need a medium of propagation. The substance that stood to waves of light as water stands to ripples on a pond was known as *luminiferous aether* (hence the expression "ripples in the aether").

But whereas we see water and the waves it carries, we see only light; we do not see an aether through which it travels. But, since science talks of many things that cannot be directly seen—for example, electrons—our failure to simply see an aether was not necessarily a problem. Electrons are allowed into science because they have properties that we can measure indirectly with the aid of instruments. Perhaps the aether could be measured indirectly with the aid of instruments. Persistent efforts failed to find an aether (and, since there were several distinct aether hypotheses, incompatible with each other, the same experiments were unable to tell which of the aethereal competitors was correct). Yet treating light as made of waves was enormously fruitful science that explained many puzzling phenomena, such as interference and diffraction effects (generated when waves, but not particles, interact).

In the minds of some scientists, this showed that the science of something useful (waves of light) established the existence of something (luminiferous aether) not merely invisible but undetectable by instruments to boot! Some distinguished scientists at the end of the nineteenth century, including Sir Oliver Lodge, whose "decoherer" was a precursor to the modern radio receiver, saw this as showing that good science could show that invisible things existed. Since Lodge was interested in spiritualism, the idea that dead people can be contacted with the help of a suitable medium and spirit guides, Lodge felt that discoveries about luminiferous aether showed that his beliefs about the further inhabitants of the realm invisible were not without justification (Powers 1982, 57–58).

Sadly, by the end of the first decade of the twentieth century, the physics of light had undergone two major shifts in thinking (resolving persistent problems that had bedeviled nineteenth-century physics). The first, initiated by Max Planck, involved the emergence of the quantum theory, according to which light was composed of discrete chunks or bundles of energy called *quanta*. The second, initiated by Albert Einstein, involved the emergence of relativity theory, which showed how to do physics without an aether. The cumulative effect was that light was no longer viewed as simply a wave phenomenon, and aether dropped out as unnecessary excess baggage. Good science didn't require that which was unobservable in principle (and possibly spiritual) after all. Methodological naturalism involves caution—often caution in the face of wishful thinking—about what is part of nature. Contrary to appearances, we had not smacked into the aether (or phlogiston, or caloric, or vital forces) after all. It is hardly surprising if more ambitious claims about supernatural intelligent design are subjected to careful scrutiny.

Modesty is a virtue in science, as well as morality. When you hear hoofbeats, think horses, not zebras. The mundane is more likely than the bizarre. Someone with the sniffles most likely has a cold. However, the bizarre cannot be excluded. Sometimes those sniffles really do herald the onset of some exotic disease. Hence, as the late Carl Sagan has advised, extraordinary claims require extraordinary evidence. Claims about the supernatural intelligent design of the universe are extraordinary claims. To become part of science, to go beyond the domain of purely religious faith, we will need extraordinary evidence for their validation. Science, running in accord with methodological naturalism, has not excluded the search for supernatural effects. Quite the reverse is true. A brief case study will help, and the one I have selected concerns the much touted beneficial medical effects of religion. In this field, there have been numerous scientific inquiries, and many publications in real science journals have been forthcoming.

Given the persistent failure of intelligent design theorists to produce any, let alone new, scientific results or to publish their "findings" in reputable journals, these published studies concerning prayer and medicine are very important. The very fact of their publication shows that scientific journals are not part of a vast, liberal, atheist, naturalist conspiracy to suppress discussion of the effects of religion from the standpoint of science.

Supernatural Science

More than seventy medical schools in the United States (out of 126) offer instruction to medical students on how to deal with the religious beliefs of their patients. At the medical school at my own university, an elective course has been offered on several occasions that is concerned with spirituality and medicine, and some researchers there have been involved in studies of the medical effects of religious belief. Sloan, Bagiella, and Powell (1999, 664) have pointed out that surveys have shown that something like 79% of the public believe spiritual faith can help people and that of the 297 physicians sampled at the 1996 meeting of the American Academy of Family Physicians, 99% were convinced that religious beliefs can heal, and 75% believed that prayers of others could promote a patient's recovery.

Some have argued on this basis that the wall of separation between medicine and religion needs to be torn down. Others have indicated that medicine of the future "will be prayer and Prozac." There is currently a trend in the United States to promote alternative healthcare modalities. Alternative medicine, ranging from homeopathy, dietary fads, and various forms of psychic or spiritual healing to certain types of chiropractic intervention and holistic medicine, enjoys enormous popular support in the United States. It is also a movement that is backed by influential lobbyists, so much so that the National Institutes of Health now has a National Center for Complementary and Alternative Medicine that disburses increasingly scarce public research funds in support of these and allied therapeutic endeavors. Given the role played by religion in the life of the nation, it is hardly surprising that the healing power of religion should find numerous advocates in the medical community.

But what are we to make of all this from the standpoint of science? The increasingly influential Templeton Foundation, in addition to supporting the work of figures prominent in the intelligent design community such as William Dembski (see Dembski, 2002, xxi), has been promoting the positive medical benefits of religion as part of an attempt to spark a constructive dialogue between science and religion. The foundation's founder, financier Sir John Templeton, has observed:

> Various research results have shown quite an extraordinary association between religious involvement, broadly considered, and likelihood of death among elderly people. At present the reason for this association is unclear. However, it is quite substantial, almost a 50% reduction in the risk of dying during follow-up and close to a 30% reduction when corrected for other known predictors of mortality. This effect on survival is equivalent in magnitude to that of not smoking versus smoking cigarettes (about seven years added to life). (2000, 109)

These are substantial claims indeed, and it will be important to examine them carefully. Koenig, Pargament, and Nielsen (1998) report some results along these lines. But are the results due to the psychology of religious belief or to a real manifestation of supernatural causes? We do not know. And such studies as we do have contain contradictory results. Thus Pargament, Koenig, Tarakeshwar, and Hahn have observed of the effects of religious struggle among medically ill elderly patients:

> Religious struggle was associated with a greater risk of mortality. Although the magnitude of the effects associated with religious struggle was relatively small (from 6% to 10% increased risk of mortality), the effects remained significant even after controlling for a number of possible confounding variables. . . . Furthermore, we were able to identify specific forms of religious struggle that were more predictive of mortality. Patients' reports that they felt alienated from or unloved by God and attributed their illness to the devil were associated with a 19% to 28% increase in risk of dying during the approximately 2-year follow-up period. (2001, 1883–1884)

Again, these results are surely interesting, but what do they tell us about the nature of the world we live in, and especially the sorts of causes that operate there?

The deduction of conclusions about the medically efficacious effects of religion from data is notoriously fraught with methodological problems. Sloan, Bagiella, and Powell (1999, 665), for example, refer to the often-cited work of Comstock and Partridge that purported to show a positive association between church attendance and health. The study was seriously flawed, alas, by its failure to control for functional capacity; "people with reduced capacity (and poorer health) were less likely to go to church." Details like this rarely get the publicity they deserve and, anyway, are of little consequence to those whose will to believe has overpowered their common sense. To examine these issues in more detail, we need to understand a bit more about the context in which matters of religion and medicine arise.

The standard medical model in scientific medicine today is something known as the bio-psycho-social model. This model embodies the idea that medical phenomena have biological causes (e.g., bacteria and viruses), psychological causes (e.g., stress and emotional disturbance), and social causes (e.g., poverty or affluence). In some quarters, especially among Christian practitioners, there has been a call for an expansion of this model so that it will become a bio-psycho-social-spiritual model. Medicine needs to be expanded to include spiritual causes. But what does this mean exactly?

We have just seen that there are two types of naturalist. In the present context, they can be contrasted as follows:

1. *Metaphysical naturalism.* There is nothing beyond nature; all causes and effects are parts of nature. There is no spiritual realm to

have any medical effects. Spiritual beliefs are beliefs, and the effects of belief are covered in the bio-psycho-social model.

2. *Methodological naturalism.* Long experience shows that all we seem to bump into in science is nature, and so all causes and effects are, with very high probability, natural, and thus the bio-psycho-social model is most probably adequate for the phenomena under analysis. Extraordinary evidence will be needed to make a case for supernatural spiritual causes in medicine, and hence an extension of the model to the bio-psycho-social-spiritual model.

The methodological naturalist is thus skeptical of claims about supernatural causes but also recognizes, since all claims in science are potentially revisable in the light of new evidence, that it is at least conceivable that all that long experience of nature has not told the whole story.

The position of the methodological naturalist may be contrasted with that of a very different kind of methodologist:

3. *Methodological supernaturalism.* Strong religious faith carries with it the view that, notwithstanding the body of established science and its experience with natural objects and their causes and effects, supernatural causes also operate in the world. The extraordinary evidence for these astonishing conclusions, prompted by faith, lies in the discovery of good evidence gathered in accord with the dictates of the best methods governing the practice of science.

The methodological supernaturalist, like the methodological naturalist, believes in gathering evidence to make a rational case for or against supernatural influences in the world. The methodological supernaturalist also recognizes that the burden of evidential responsibility rests firmly on his or her shoulders. Perhaps firm religious convictions incline methodological supernaturalists to undertake these sorts of scientific studies, but the position is at least tenable with its recognition of the evidential burden. So, with these three different positions available to thoughtful people, what have we actually learned from studies of spirituality in medicine?

The results of studies are generally very messy and hard to interpret. This should hardly surprise us, since epidemiological studies about natural causes, let alone supernatural causes, are hard to conduct (see, for example, Knapp and Miller 1992). Moreover, Sloan,

Bagiella, and Powell (1999) have reviewed the literature on spirituality in medicine and have found studies to be plagued with a whole host of methodological problems. But this does not mean they are all so plagued. And it does not mean that there are no results out there worthy of further examination.

For example, experimental studies that take care to control for extraneous causal influences would be helpful, and such experiments have been conducted. A very important study often cited by advocates of the supernatural is the famous Harris study. This study is a published report of an attempt to perform a controlled, randomized, double-blind prospective study of the effects of remote intercessionary prayer on patients in a coronary care unit (Harris et al. 1999). The study concerned 990 patients in a university-affiliated hospital in Kansas City. The study received extensive, uncritical publicity in the news media.

There is a noble intellectual precedent for conducting such studies, coming from none other than Darwin's relative, Sir Francis Galton (1822–1911). Galton, who was one of several towering figures in Victorian science and a pioneer in the application of statistical methods to scientific problems, believed that anything that could be measured was a legitimate subject for scientific inquiry. To this end, he even proposed a scientific statistical study of the effectiveness of prayer (Gould 1981, 75).

It is worth pointing out before proceeding further that a statistical study of the kind we are going to examine might show prayer to be medically effective—it might establish the existence of a statistical correlation. But such a study will not explain *why* prayer is medically effective. We must never confuse causation with correlation. (In children, arm length is positively correlated with general levels of cognitive development—but this is because older kids have longer arms. Only a fool would try stretching his child's arms with a view to accelerating cognitive development.) It is generally easy to find correlations. Figuring out causation is usually much harder.

In the study we are examining here, the patients were divided into a prayer group and a usual care group. Members of the prayer group were known to the prayer-providers by their first names. The prayer-providers were unknown to the patients. Members of the prayer group received daily prayer for four weeks. Based on the scoring system employed in the study, the prayer group did about 10% better than the usual care group. The study was widely reported in the media and published in the prestigious *Archives of Internal Medicine*. The study concluded that prayer may be a useful adjunct to standard care. With the rising cost of medicine, this looks like a good faith-based initiative. Given the extraordinary nature of the claim, if validated (as all claims in science must be), it would surely be one of the great discoveries of the twentieth century.

In science, it is a standard practice to examine very carefully the data, the way it was analyzed, and the way it was gathered. These examinations are particularly vigorous if a claim is a truly extraordinary claim. For example, the claim, by Pons and Fleischmann, that cold fusion had been achieved in the laboratory—a truly astonishing claim about nature—fell from grace as numerous methodological and data-analysis problems came to light. Issues about methodology and data analysis are among the first things that scientists look at when examining published reports of experimental results. This scrutiny is simply good science in action and not a manifestation, in the present case, of antireligion bias.

The first observation is that the study itself made clear that there was no significant difference between the prayer group and the usual care group with respect to speed of recovery. The differences were with respect to a scoring system used in the study. The scoring system used in the study lacked independent scientific validation. It represented a choice by the researchers to stick numerical values on events; whether they reflect medically significant valuations is not scientifically established. This feature of the study led to criticisms. Thus Sloan and Bagiella have observed:

> On both unweighted and weighted scales, the prayer group showed a slightly but significantly better clinical course (i.e., lower scores) than the control group. The unweighted score is completely meaningless . . . a patient who dies in the cardiac care unit has a lower unweighted score (1 event) than one who requires antibiotics, arterial monitoring and antianginal agents (3 events). The significance of the group differences on the weighted scale assumes it has construct validity (e.g., need for an electrophysiological study (3 points) is 3 times as bad as the need for antibiotics (1 point). . . . This is by no means clear. (2000, 1870)

But this concern about use of a seemingly arbitrary scoring index is a concern that plagues many studies that have nothing to do with spirituality. Observing this problem here is not a manifestation of antireligion bias, and it is an issue that needs further analysis. Looking at the published data that accompanied this study, Dudley Duncan (personal communication) has pointed out to me that there were issues about how significant, statistically, the results really were. But did the study show absolutely nothing? Methodological naturalists and methodological supernaturalists will want to know more, and there is certainly material here to pique one's natural curiosity.

For example, in the study by Harris et al., we learn (1999, 2273) that of the 1,013 patients enrolled in the study, 484 were originally assigned to the prayer group and 529 to the usual care group. Now because it took a full day to get prayer up and running, patients who spent less than twenty-four hours in the coronary care unit were dropped from the study, leaving 466 (a loss of 18 patients in the prayer group) and 524 (a loss of 5 patients from the usual care group). This difference between the two groups is statistically significant. (The chi-squared p value for this difference was <0.001; i.e., it could have happened by chance with a probability of less than one in a thousand. The p value reported for the main result of the study, by contrast, was <0.04; i.e., it could have happened by chance with a probability less than 4 in 100. See Nicholas Humphrey in *Skeptical Inquirer*, www.csicop.org/si/2000-05/prayer.html.)

Several hypotheses suggest themselves at this juncture. If we dismiss the difference between the prayer group and the usual care group with respect to those dropped from the study as occurring just by chance, here are some possibilities:

(H1) People got better just by being assigned to the prayer group. It was enough that they would have been prayed for, even though they were not, that made them well. (This tongue-in-cheek suggestion was made by Nicholas Humphrey.)

(H2) Perhaps the people who were dropped didn't get better but died instead. (Harris himself apparently made this suggestion to Nicholas Humphrey.) Maybe so, but now being assigned to the prayer group might be interpreted as having deadly consequences.

(H3) There was a hidden bias in the study that assigned healthier patients to the prayer group than to the usual care group. This bias could then explain the better scores by patients in the prayer group without the need for supernatural

> intervention. (We do not have to know this was the case; it is enough that steps were evidently not taken by the authors of the study to rule it out. The burden of proof lies on their shoulders, not ours.)

Without knowing more about the data than this (the raw data, alas, is apparently not available for study), methodological modesty seems prudent. We clearly hear hoofbeats, but it is most likely horses and not zebras. There is probably some hidden bias in the assignment of patients to the two groups that needs to be rooted out and controlled for. Until this issue is dealt with, we certainly have no solid ground to leap to more potent conclusions about the prayer study and the mysterious effects of prayer at a distance.

In this regard, the eighteenth-century philosopher David Hume, whom we met in chapter 1, had occasion to remark: "When anyone tells me he saw a dead man restored to life, I immediately consider with myself, whether it be more probable that this person should either deceive or be deceived, or that the fact, which he relates, should have really happened. I weigh the one miracle against the other" (1975, 116). Hume is not saying that the extraordinary claim is false but that without additional evidence we need to be cautious, lest we acquiesce in the extraordinary too readily. Without knowing more, H3 seems a reasonable position to adopt.

So what we learn from the data in these studies about spirituality in medicine is that the studies themselves are a long way from providing us with unambiguous evidence of the healthful effects of religion. They are far from conclusive. So as yet, we cannot currently find clear support for a simple conclusion: be religious; it is good for you. Even less can we find clear evidence of supernatural causation. Should we come across a truly well-constructed study that provided unambiguous evidence of the effectiveness of intercessionary prayer, for example, we would want to investigate much more closely hypotheses about causation. It does not simply follow from the observation that prayer is effective that there has been supernatural intervention. I, for one, await with interest the results of such a study, because, if this field of research is not to be hopelessly tarnished by methodologically flawed studies, good studies, perhaps with the help of those more skillful in bioinformatics than is apparent in many of the studies that have been reviewed to date, would be most useful.

As for the current studies of prayer in medicine, I think a fair verdict is "case not made." Much of the critical discussion and defense of these studies hinges on arcane topics like statistical significance and research design. In the uncritical exploitation of such studies, we find Christian apologists speaking confidently about the meaning and implications of statistical tests of significance, which, it is clear, the investigators themselves understand none too well.

But what of the different theorists we met previously? The metaphysical naturalist might say that such effects as there are can be accounted for in terms of belief, attitude, social factors, and bad methodology. The methodological naturalist will be inclined to agree. The evidence presented simply does not warrant an expansion of the medical model to a bio-psycho-social-spiritual model. And in the particular cases examined here, presumably an intellectually honest methodological supernaturalist will agree as well. There may be interesting effects, but these do not clearly and distinctly herald the introduction of supernatural causation and hence an extension of the medical model to include a non-biological, non-psychological, non-social, spiritual dimension. We do not yet have the extraordinary evidence to support supernatural causation.

What this means is that Dembski's claim, stated in the previous section, to the effect that methodological naturalism is functionally equivalent to metaphysical naturalism needs to be understood carefully. In the case of the studies we have just mentioned, the metaphysical naturalist, the methodological naturalist, and the methodological supernaturalist might well agree with each other. But the agreement between the methodologists here derives from the fact that in the studies we looked at, there is no unambiguous, compelling evidence in support of supernatural causation. This does not mean that such evidence might not one day be forthcoming. If it did, the methodological naturalist might well part company with the metaphysical naturalist. But such evidence would have to be extraordinary evidence to support such an extraordinary claim.

So methodological and metaphysical naturalism are nevertheless different positions. They differ with respect to the practice of science. On the former view but not the latter, supernatural hypotheses can be entertained. I suspect that the real reason Dembski wants to run the two distinct theses together as being functionally equivalent, in the context of claims about supernatural intelligent design, is that accompanying methodological naturalism is a demand for the production of clear, unambiguous evidence, and it would be as well to ensure that this demand for such evidence is conflated with a purely philosophical theory according to which none is required right at the outset.

But wait. I have just said that intelligent design claims are really claims about the supernatural. Is this a fair characterization? As we will see, it is indeed fair. But watch out for the thin end of the wedge, for there you will find much evasion and obfuscation.

Natural and Supernatural Intelligent Design

The slipperiness to be found at the thin end of the wedge is nicely captured by the following remarks by Dembski:

> Science, we are told, studies natural causes, whereas to introduce God is to invoke supernatural causes. This is the wrong contrast. The proper contrast is between *natural causes* on the one hand and *intelligent causes* on the other. Intelligent causes can do things natural causes cannot. . . . Whether an intelligent cause operates within or outside nature (i.e., is respectively natural or supernatural) is a separate question entirely from whether an intelligent cause has operated. (1999, 105)

These are fair distinctions. Lots of us—myself included—are interested in intelligent design of artifacts in ways that have no supernatural implications. (My own interest is in antique surgical instruments.) Of course, there is evidence of natural intelligent design of human artifacts. Who would be silly enough to deny it?

The trouble here is that the evidence of natural, human design of artifacts is utterly irrelevant to the issue as to whether there is supernatural intelligent design of the universe and its contents. The proper contrast, contrary to Dembski, is between natural causes (including those behind the production of human artifacts) and supernatural causes. Make no mistake: This debate is all about supernatural intelligent design by the God spoken of in the Christian religion.

Interestingly, intelligent design theorist Michael Behe, in a reply to some of the objections Karl Joplin and I had developed in response his ideas, took us to task for attributing to him an interest in supernatural creation. Like Dembski, he likes to give the appearance of treading carefully. Behe directed us to the thin end of the wedge: "The core claim of intelligent design theory is quite limited. *It says nothing directly about how biological design was produced, who the designer was, whether there has been common descent, or other such questions. Those can be addressed separately*. It says only that design can be empirically detected in observable features of physical systems" (2001a, 15, my italics).

But perhaps, at the very thinnest end of the wedge, these other issues about design cannot be addressed at all. In the same reply, Behe approvingly quoted Kenneth Miller (himself a believer in cosmological intelligent design) to the effect that a clever God might do the designing in ways "scientifically undetectable to us" (2001a, 15). Behe added, "I have no reason to object to that as a route to irreducibly complex systems . . . and that while we may be unable to discern the means by which design is effected, the resultant design itself may be detected in the structure of an irreducibly complex system" (ibid.). (We will meet *irreducible complexity* in detail in the next chapter.)

Behe (1996, 249) even mentions naturalistic hypotheses about design. Perhaps it was done by aliens from a galaxy far, far away or by a time-traveling biochemist (see also Behe 2001b, 101). Perhaps because appeals to natural designers do not really solve the problem but only postpone a confrontation with it (who designed the alien designers of life on earth?), Behe added, "Most people, like me, will find these scenarios entirely unsatisfactory, but they are available for those who wish to avoid unpleasant theological implications" (1996, 249). Behe thinks reasonable people will conclude that the designer "is beyond nature" (2001b, 101), that is, supernatural.

In a similar vein, Dembski (1999, 261–263; 2001a, 7) has responded to criticisms directed against intelligent design hypotheses to the effect that the organisms we see in nature show evidence of either bad design or suboptimal design. The issues here go beyond the old joke that a giraffe looks like a horse designed by committee. For example, getting around on our hind legs causes many problems from knee and ankle trouble to lower back pain to hemorrhoids (so, thank goodness, we have intelligently designed soothing ointments).

To an evolutionary biologist, the appearance of poor design is evidence of the operation of a bungling, unintelligent trial-and-error evolutionary process that has resulted in suboptimal anatomical structures. Biologists point to these sorts of examples because they seem hard to account for if the intelligent design was due to an all-knowing, all-good, all-powerful designer, supernatural or otherwise. And this was precisely the sort of designer who has appeared in religious objections to evolution. The point is that if these defective structures were the result of design, then the designer must presumably have been drunk, stupid, or both!

Once again, at the thin end of the wedge, Dembski has taken these critics of intelligent design to task by making what looks like a reasonable distinction. He observes, "The confusion centered on what the adjective intelligent is doing in the phrase 'intelligent design.' 'Intelligent' can mean nothing more than being the result of an intelligent agent, even one who acts stupidly. On the other hand it can mean that an intelligent agent acted with skill, mastery and *éclat*" (2001a, 7). It should be pointed out, however, that while Dembski is technically correct, neither he nor anyone else in this debate is interested in dumb design. This is merely a sop to forestall a confrontation with the serious issue of the overwhelming evidence for suboptimal design in nature.

In a similar vein, responding to the version of the problem of evil that led Darwin to be suspicious of claims about intelligent design, Dembski has observed:

> Critics who invoke the problem of evil against intelligent design have *left science behind and are engaging in philosophy and theology*.
>
> Design by intelligent agency does not preclude evil. A torture chamber replete with implements of torture is designed, and the evil of its designer does nothing to undercut the torture chamber's design. The existence of design is distinct from the morality, esthetics, goodness, optimality or perfection of design. (2001a, 9–10, my italics)

Once again, the statement is technically correct but irrelevant. The motivation behind the intelligent design movement is to justify the claim that there is evidence for a supernatural designer indistinguishable from the God of Christianity—not some idolatrous Hindu deity, not some incompetent, stupid, supernatural bungler, and not some evil manufacturer of torture devices. Dembski and his friends know this as well as I do.

We get closer to the real issues at hand when Dembski discusses the criticism that the modern intelligent design movement is about the reintroduction of magic into science. He thinks opposition to intelligent design theory springs from a fear that if science admits a transcendent designer, this will destroy science, reintroducing into our account of the world "all sorts of magical, superstitious, and occult entities that modern science has thankfully banished for our understanding of the world" (2001a, 18).

In responding to Robert Pennock's charge that intelligent design theory involves recourse to magic and the supernatural, Dembski observes, "Intelligent design is detectable; we do in fact detect it; we have reliable methods for detecting it; and its detection involves no recourse to the supernatural" (2001a, 19). Once again Dembski has made a correct but irrelevant observation. The issue is not whether we can detect design or even whether our methods of detection involve recourse to the supernatural. The problem lies in Dembski's unwarranted leap to the conclusion of supernatural design using his methods for the mere detection of design. To appreciate this last observation, we must leave the thin end of the wedge for the fat end.

The Fat End of the Wedge

At the thin end of the wedge, intelligent design theory seems to be so innocuous as to be hardly worth the effort of refutation or argument. But the thin end of the wedge has been carefully crafted to get a hearing from audiences wary of the excesses of creationist silliness. Not for nothing is intelligent design theory known as stealth creationism. Once it is in place, however, the wedge can be hammered home, and what we really need to know is what is lurking at the fat end of the wedge, for that is where the real business of intelligent design theory is to be found.

Once again, William Dembski, the same proponent of intelligent design theory who would have us believe that it is merely concerned with intelligence—maybe natural, maybe dumb, maybe evil—has also claimed that Christian theology is to be the queen of the sciences:

> I want to consider what it means to assert the preeminence of Christian theology among the disciplines and particularly among the sciences. My thesis is that all disciplines find their completion in Christ and cannot be properly understood apart from Christ. If we take seriously the word-flesh Christology of Chalcedon (i.e., the doctrine

> that Christ is fully human and fully divine) and view Christ as the *telos* toward which God is drawing the whole of creation, then any view of the sciences that leaves Christ out of the picture must be seen as fundamentally deficient. (1999, 206)

He adds:

> So, too, Christology tells us that the conceptual soundness of a scientific theory cannot be maintained apart from Christ. Christ is the light and the life of the world. All things were created by him and for him. Christ defines humanity, the world and its destiny. It follows that a scientist in trying to understand some aspect of the world, is in the first instance concerned with that aspect as it relates to Christ—and this is true regardless of whether the scientist acknowledges Christ. (1999, 209)

Later in the same volume, we learn that creation proceeds through the spoken word of God (232), that the reality of a blade of grass lies in its ability to communicate with God (232), that because of this, communication theory is the fundamental science (233), and that the fine-tuning of the universe and irreducible biochemical complexity are instances of information "inputted into the universe by God at its creation" (233). (On the dust jacket of this book, Dembski is described as "the Isaac Newton of information theory." High praise indeed, given this last claim.) We also learn that creation took six days and left God tired (234). To quote Dembski's own words back at him, we have clearly "left science behind and are engaging in philosophy and theology."

We are a long way from dumb, bungling, possibly evil, natural designers here. If there is any doubt about this, consider Dembski's own judgment concerning intelligent design theory: "The world is a mirror representing the divine life. The mechanical philosophy was ever blind to this fact. Intelligent design, on the other hand readily embraces the sacramental nature of physical reality. Indeed, intelligent design is just the Logos theology of John's Gospel restated in the idiom of information theory" (2001b, 192). What we have is a cluster of philosophically bloated, theologically pretentious, and evidentially unwarranted claims about supernatural design. When proponents of intelligent design theory deny that they are trying to teach religion dressed up as science, they frankly violate their own religion's prohibitions against telling lies. If policy makers and educators are deceived in this regard, it can only be because they have not examined the claims of intelligent design theory in any detail—a state of ignorance readily exploited by the proponents of intelligent design.

But Dembski has told us there is evidence. It would be wrong to dismiss it out of hand. As evidence of intelligent design, he has cited irreducible biochemical complexity. That is the subject of the next chapter. He has also cited the fine-tuning of the universe. That will be examined in the chapter that follows. A discussion of the disturbing social implications of the wedge strategy will be undertaken in the conclusion.

5

The Biochemical Case for Intelligent Design

I think it is fair to say that *Darwin's Black Box: The Biochemical Challenge to Evolution* (1996), by creationist biochemist Michael Behe, is without doubt the most influential of the recent books written in support of intelligent design. On its face, it is an attempt to articulate a principled argument from the study of nature to the conclusion that nature contains features that require intelligent design. The argument is derived from a long line of arguments (reviewed in chapter 1) that culminated in Paley's version of the design argument. At the thin end of the wedge, much of the intelligent design movement can be thought of as footnotes to Behe.

Darwin showed how Paley's argument from design failed to establish its conclusion, and Behe's response has been to claim that there were things Darwin was unaware of but which show that in biology there is evidence of intelligent design. The evidence concerns a type of biochemical complexity that Behe has termed *irreducible complexity*. Behe's claim is that systems exhibiting irreducible complexity do not admit of a plausible Darwinian explanation and are better explained as the fruits of supernatural intelligent design.

In this chapter, I will explain why I think Behe is profoundly mistaken. Along the way, we will discover that Behe's puzzle about irreducible complexity, for which he has now become famous, was posed independently a decade earlier by another biochemist, A. G. Cairns-Smith (1986). Given that Behe (1996, ch. 8) did an extensive survey of the literature concerning evolution in biochemistry, it is curious that he missed Cairns-Smith's little volume. No matter. Since Cairns-Smith also sowed the seeds of a purely naturalistic explanation of irreducible complexity—seeds that will be allowed to germinate in this chapter—we shall endeavor here to give Cairns-Smith his due.

Irreducible Complexity

We will begin with a discussion of Behe's characterization of the nature of the problem. Behe initially characterized an irreducibly complex system as follows: "a single system which is composed of several well-matched, interacting parts that contribute to the basic function, and where the removal of any one of the parts causes the system to effectively cease functioning" (1996, 39). In a later work, Behe put it this way: "But what type of biological system could not be formed by 'numerous, successive, slight modifications'? A system that is *irreducibly complex*. Irreducible complexity is just a fancy phrase I use to mean a single system that is composed of several interacting parts, where removal of any one of the parts causes the system to cease functioning" (2001b, 93).

Notice that in this latter formulation, the requirement that the parts of irreducibly complex systems be well-matched has been dropped. This will be a source of controversy later in this chapter. Nevertheless, such irreducibly complex systems are alleged to pose a difficulty for Darwinian theory because: "An irreducibly complex system cannot be produced directly (that is, by continuously improving the initial function, which continues to work by the same mechanism) by slight, successive modifications of a precursor system, because any precursor to an irreducibly complex system that is missing a part is by definition nonfunctional" (1996, 39). Later, Behe put it this way: "An irreducibly complex biological system, if there is such a thing, would be a powerful challenge to Darwinian evolution. Since natural selection can only choose systems that are already working, then, if a biological system cannot be produced gradually, it would have to arise as an integrated unit for natural selection to have anything to act on" (2001b, 94). Put this way, the problem is indeed like the old puzzle about the eye: How could something whose function (and value to the organism) requires the coordinated action of many parts arise gradually from the action of unguided natural processes that have no view to future utility or purpose?

Is it absolutely impossible for Darwinian mechanisms to explain irreducible complexity? Behe observes:

> Demonstration that a system is irreducibly complex is *not* a proof that there is absolutely no gradual route to its production. Although an irreducibly complex system cannot be produced directly, one can't definitively rule out the possibility of an indirect circuitous route. However, as the complexity of an interacting system increases, the likelihood of such an indirect route drops precipitously. (2001b, 94)

Given the weight that irreducible complexity carries with proponents of intelligent design, this is a significant admission. Whether indirect pathways to irreducible complexity are unlikely is nowhere demonstrated by Behe. So we must pay careful attention in what follows to unsubstantiated claims about lack of likelihood, lest they turn out to be mere indicators of ignorance or prejudice on the part of Behe, and not real measures of what can and cannot happen in the world.

A decade before Behe laid out his biochemical challenge to evolution in terms of irreducible complexity, another biochemist, A. G. Cairns-Smith, had laid out the same problem, albeit in slightly different language. Cairns-Smith observed:

> The bit that is not so clear about the eye—and a favorite challenge to Darwin—is how its components evolved when the whole machine will only work when all the components are in place and working.
>
> Not that this problem is peculiar to the eye. Organisms are full of such machinery, and it is a widely held view that this appearance of having been designed is *the* key feature of living things. . . . How can a complex collaboration between components evolve in small steps? (1986, 58)

Cairns-Smith, comparing the pathways of the central biochemistry of organisms with stone arches in which all the components of the arch depend on each other, observed:

> Nowhere is a collaboration of components tighter than in central biochemistry. Pull out a molecule—any molecule . . . you will find that every molecule is required in some way or other by every other

> molecule.... Nothing can be touched or the whole edifice will collapse. Looking at the structure of interdependencies in central biochemistry it is not at all difficult to see why central biochemistry is now so fixed and has been for so long. The difficult question is how such a complexity of arching evolved stone by stone. (1986, 60)

Behe's puzzle about irreducible complexity was evidently being bounced around in the popular biochemistry literature at least ten years before he wrote about it.

More than this, Cairns-Smith even speculated about the possibility of intelligent design. Indeed, chapter 3 of his book is devoted to an exploration of the problems confronting a would-be intelligent designer of an *E. coli* bacterium—a discussion that brings to the fore the importance of thermodynamic considerations. Later in the same book, Cairns-Smith remarked: "We may make a machine by first designing it, then drawing up a list of components that will be needed, then acquiring the components, and then building the machine. But that can never be the way evolution works. It has no plan. It has no view of the finished system. It would not know in advance which pieces would be relevant.... *It is the whole machine that makes sense of its components*" (1986, 39). The very difference between evolutionary explanations and explanations in terms of intelligent causes is already there in Cairns-Smith's work. But Cairns-Smith wisely rejected appeals to the supernatural simply to fill in gaps in our knowledge:

> It is a sterile stratagem to insert miracles to bridge the unknown. Soluble problems often seem to be baffling to begin with. Who would have thought a thousand years ago that the size of an atom or the age of the Earth would ever be discovered?... It is silly to say that because we cannot see a natural explanation for a phenomenon that we must look for a supernatural explanation. (*It is usually silly anyway*.) With so many past scientific puzzles now cleared up there have to be very clear reasons not to presume natural causes. (1986, 6, my italics)

Cairns-Smith is clearly a methodological and not a metaphysical naturalist. He does not exclude the supernatural but thinks you need more than current ignorance to support the need for explanations involving supernatural intervention. Later in this chapter, we will employ ideas derived from Cairns-Smith's work to lay this problem of irreducible complexity to rest.

Information Theory and Irreducible Complexity

Behe's arguments, original or not, are the linchpin of the intelligent design movement. Behe's central argument is that biochemical systems exhibit irreducible complexity that cannot have evolved as the result of unintelligent natural causes and must have appeared as the result of intelligent causes. (We saw in the last chapter that Behe's predilections run to supernatural intelligent design, and hence to supernatural causation.) In chapter 3, we saw that other members of the intelligent design movement have tried to create criteria by means of which we might be able to recognize intelligent design in patterns, processes, and structures that we can observe in nature.

William Dembski in particular has devoted much effort to the construction of what he terms the *complexity-specification criterion*, which is designed to help us detect design (see Dembski 1999, 2001b). The criterion is important in the present context because a crucial application singles out Behe's irreducibly complex systems as the fruits of intelligent design. Dembski observes:

> The connection between Behe's notion of irreducible complexity and my complexity-specification criterion is now straightforward. The irreducibly complex systems Behe considers require numerous components specifically adapted to each other and each necessary for function. On any formal complexity-theoretic analysis, they are complex in the sense required by the complexity-specification criterion. Moreover, in virtue of their function, these systems embody patterns independent of the actual living systems. Hence these systems are also specified in the sense required by the complexity-specification criterion. (1999, 149)

Because the complexity-specification criterion underlies the design inferences that intelligent design theorists wish to foist on us, and because the criterion singles out irreducibly complex biochemical systems as intelligently designed systems, a critical examination of Behe's claims about irreducible complexity will have broad implications for the intelligent design movement as a whole, especially if it should turn out that irreducibly complex systems could plausibly have evolved without intelligent design, for this would show that not only is there something defective in the account intelligent design theorists offer of biological phenomena but also there is something profoundly deceptive about the methods they use to detect intelligent design.

Mechanical Metaphors

In chapter 1, we saw how mechanical metaphors played a crucial role in early scientific reasoning about design. This tradition is continued in the work of Michael Behe. To explain irreducible complexity, Behe has employed the mechanical metaphor of the spring-loaded mousetrap. The mousetrap is to Behe what the well-designed pocket watch was to William Paley. Here is Behe explaining his mechanical analogy:

> The mousetraps that my family uses consist of a number of parts. There are: (1) a flat wooden platform to act as a base; (2) a metal hammer, which does the actual job of crushing the little mouse; (3) a spring with extended ends to press against the platform and the hammer when the trap is charged; (4) a sensitive catch that releases when slight pressure is applied; and (5) a metal bar that connects to the catch and holds the hammer back when the trap is charged. Now you can't catch a mouse with just a platform, add a spring and catch a few more mice, add a holding bar and catch a few more. All the pieces of the mousetrap have to be in place before you catch any mouse. Therefore the mousetrap is irreducibly complex. (2001b, 93–94)

Whether the metaphor is that of the mechanical pocket watch, with its many finely crafted moving parts, or the mass-produced mousetrap, it will not go amiss to discuss the metaphor itself before proceeding to biochemistry. While ingenious tinkerers have tried to show how you can have a functional mousetrap with fewer than the five parts Behe has listed, I am not convinced that this is where the problems with the metaphor ultimately lie.

The reason for this is that scientists themselves have employed mechanical metaphors repeatedly in their explanations of how biological systems work. Here is Bruce Alberts, president of the National Academy of Sciences:

> The entire cell can be viewed as a factory that contains an elaborate network of interlocking assembly lines, each of which is composed of large protein machines. . . . Why do we call the large protein assemblies that underlie cell function *machines*? Precisely because, like

> machines invented by humans to deal efficiently with the macroscopic world, these protein assemblies contain highly coordinated moving parts. (quoted in Dembski 1999, 146–147)

For Alberts, the mechanical metaphors are a *façon de parler*, a convenient way of talking for the purposes of explanation that should not be taken too literally. Behe takes mechanical metaphors more literally as being indicative of the nature of reality. Machines are designed systems, and biochemical systems, by comparison, don't just look as if they are designed, they are designed. Behe cannot be faulted for drawing on our cultural experiences with mechanical contrivances of varying degrees of complexity. It is something all of us do. There are more important issues, however, and we must now examine them.

Intelligent Design and the Origins of Artifacts

At the thin end of the wedge, it is claimed that there are features of the world that stand as evidence of intelligent design, even though we do not know who the designer was, how design was accomplished, or to what end. This, however, is the very territory where we must proceed carefully with reasoning based on mechanical metaphors. The reason can be stated simply. For ordinary objects like mousetraps and watches, we can safely infer that they are designed without having any knowledge of their designer. Why? Because we all know mousetraps and watches are *human* artifacts, and by definition human artifacts have human designers, even if their identities are obscure. This design inference, confined to the domain of human artifacts, is quite acceptable (and largely uninteresting).

To see why, imagine you are strolling across a heath and you stumble upon an old pocket watch. Knowing antecedently that watches are human artifacts, you know it is the fruit of intelligent human design, though being ignorant of the history of watchmaking, you know not how or by whom. The hand of the human designer is a mystery. Your interest may end just here. It is a nice watch, and whoever made it did a good job. But suppose your curiosity is aroused by this discovery. You will then want answers to what Karl Joplin and I have termed the *intelligent design questions*: (1) What is it? (2) Who made it? (3) When was it made? (4) How was it made? (5) For what purpose was it made?

All of these questions can be asked about the alleged fruits of design of any kind, be they watches, crop circles, stone tools, signals from outer space, or biochemical pathways.

Getting back now to the watch you found on a lonely heath, research in libraries and museums might lead you to the conclusion that the watch was a quarter repeater, made by Abraham-Louis Breguet, in 1814, in Paris, France. Studies of the watch itself reveal that it is a quarter repeater on gongs with push pendant, ruby cylinder escapement, gilt finish, serial number 2371. Further studies reveal a silver engine turned dial with the Breguet secret signatures. Research into the records of Breguet Atelier lists twenty artisans who were paid to make parts for this watch. Finally, through careful investigations into the history of technology, you are able to uncover the nature of the materials and methods employed by Breguet.

Suppose now you are strolling through a laboratory, guided perhaps by Bruce Alberts, whom we met earlier, and you come across a human cell. Microscopic examination reveals that it, like the watch, is a complex system with many interacting parts. But there is a crucial difference. Because for all of the use of mechanical metaphors to make sense of intracellular processes and structures, it is not known antecedently that the cell is the fruit of intelligent design. Here is the crucial point. Just saying that biochemical systems are like familiar human artifacts such as mousetraps or watches, or behave as if they were machines or designed artifacts, does not make them into artifacts or machines possibly of alien or supernatural origin. And in fact, from the standpoint of complexity of organization and dynamic interaction, they are utterly unlike anything in everyday human experience of designed artifacts.

No metaphor is perfect. When we say X is analogous to Y, where X is something we are struggling to understand and Y is something we already understand, X is usually like Y in some respects and not others. For this reason, metaphors, while they can help us grasp the unfamiliar, can also be misleading. Mechanical metaphors and analogies may render biochemical systems familiar and more tractable to thought, but they do not transform these systems into artifacts.

The claim that biochemical systems are intelligently designed systems is a claim that needs evidential justification. The very issue of design itself—an issue we did not have to confront for watches and other things antecedently known to be humanly designed—is an issue that must be confronted here. About the only way to provide hard evidence that systems whose intelligent design is in doubt are in fact fruits of intelligent design after all is to provide a lot of high-quality evidence to support specific and unambiguous answers to the intelligent design questions. Settling the very issue of the intelligent design of biochemical systems is thus inextricably intertwined with the provision of evidentially grounded answers to the intelligent design questions, just because it is not known antecedently that biochemical systems result from deliberate design by natural or supernatural agent(s).

Since Behe has made much of the mousetrap metaphor, it is worth noting that the case is very different where mousetraps are involved. We know rather a lot about the origin of mousetraps. Bungling trial-and-error methods, with selective retention of successful variants for further modification, have played an important role. Since the U.S. Patent Office opened in 1838, it has granted more than 4,400 mousetrap patents. Currently, about forty new patents are issued each year. Ten times that many patents are turned away, mostly because they are not minimally viable. The patent office mousetrap taxonomy recognizes thirty-nine subclasses that include "Impalers," "Smiters," "Swinging Strikers," "Choking or Squeezing," "Constricting Noose," and "Electrocuting and Explosive" (Hope 1996, 92).

Devices that kill mice by hitting them have a long and interesting evolutionary history (see Hornell 1940). The spring-loaded trap discussed by Behe appeared in the 1890s and was patented in 1903 (No. 744379) by John Mast, a Pennsylvania coleslaw manufacturer with a serious rodent problem. The spring-loaded trap did not result from design and creation from nothing—a secular miracle in Pennsylvania—for John Mast had studied existing mousetrap patents and had borrowed from five or six of them (thus showing the importance of horizontal information transfers) before filing his own patent application in October 1899 (Hope, 1996, 94). Behe's mousetrap is in fact a technological hybrid, descended with modification from earlier traps in a complex, historical, evolutionary process. While the mousetrap is intelligently designed, it didn't appear by a magical, ahistorical process of special creation, the details of which are forever hidden from public view!

Moving now to biochemistry, it is already possible for humans to intelligently design human proteins, and even novel proteins that have never been seen before. Some recent experiments have shown that more complex structures, such as the poliovirus, can be made in the laboratory. Perhaps one day it will be possible for a clever biochemist to intelligently design a functioning cell indistinguishable from a human cell. Even if this were so, it would not demonstrate that *our* cells were the results of human intelligent design, still less of alien or supernatural design.

Moreover, whether a given cell under study was extracted from a human subject or intelligently designed in the laboratory to look just like it is a matter that could be settled only through analysis of its causal history. This information will not be discernible simply by inspecting the cell itself. There is nothing analogous to the Breguet secret signatures to be found there. The judgment would be in favor of human intelligent design, if the trail led back to the laboratory, to identifiable human designers with the biochemical and biological wherewithal to accomplish the feat. The judgment would favor intelligent human design precisely if the design questions could be appropriately answered and justified. Then and only then would we have a good account of the intelligent origins of the cell in question. Supernatural accounts of the intelligent origins of cells are utterly vacant until these issues are confronted through the provision of high-quality evidence.

And we have been mistaken about the very signs of intelligent design before. When Anthony Hewish and Jocelyn Bell discovered the first pulsar in 1967 (an astronomical object emitting bursts of microwaves every 1.33730109 seconds), they wondered whether it was a beacon—a sort of cosmic lighthouse—signaling the existence of an alien intelligence. The objects were even referred to as "Little Green Men" (Dixon 1980, 402). The intelligent design theory of pulsars dropped from sight, partly because plausible answers to the intelligent design questions were not forthcoming (e.g., Why broadcast in such a messy part of the electromagnetic spectrum? Why expend so much energy? And, as more objects were discovered, why send signals to Earth from so many different places using similar frequencies?). And also because it was discovered that there were natural, unintelligent explanations for the same phenomenon—rapidly rotating neutron stars (Asimov 1971, 306).

The objections and worries I have been considering so far come down to this: Intelligent design theorists are very interested in the origins of things. Things in nature, such as biochemical systems, are said to have their origins in intelligent design. But at this point their curiosity about issues of origins mysteriously comes to an end. How did the design originate? By what processes was it effected? These issues are left unanswered as mysteries beyond the ken of design theorists. But it is only by looking at origins of design, by providing evidentially grounded answers to the design questions, that we will be able to get beyond the claim that something appears as if it was designed to the desired conclusion that it was, as a matter of fact, designed.

A special case here concerns the supernatural designer itself. Responding to objections in which critics ask about the design of the intelligent designer itself, Dembski has observed:

> The who-designed-the-designer question invites a regress that is readily declined. The reason this regress can be declined is because such a regress arises whenever scientists introduce a novel theoretical entity. For instance, when Ludwig Boltzmann introduced his kinetic theory of heat back in the late 1800s and invoked the motion of unobservable particles (what we now call atoms and molecules) to explain heat, one might just as well have argued that such unobservable particles do not explain anything because they themselves need to be explained. (2002, 354)

The other side of this tale from the history of science (with the devil lurking in the details) is that there was a century-long controversy about the very existence of atoms and molecules, which was initiated by Dalton's use of the atomic hypothesis in chemistry in 1808 and Avogadro's differentiation between atoms and molecules in 1811.

This debate continued throughout the nineteenth century and culminated in a bitter dispute between Boltzmann and Mach over evidence for the very existence of unobservable atoms and molecules. This debate was not settled until after the turn of the twentieth century, when evidence began to accumulate from many independent quarters, such as the study of radioactive decay, the quantum theory, and early experimental efforts to determine a value for the Avogadro number (see Mason 1962, ch. 39). Independent evidential warrant, not Boltzmann's say-so or even mere explanatory utility, was what settled the issue in favor of atoms and molecules.

In the case of supernatural intelligent designers of unknown constitution using unknown methods and materials to unknown ends, we have neither independent evidential warrant nor even mere explanatory utility. Dembski's suggestion that we stop and content ourselves with the progress we have made is utterly fatuous. The question, Dembski advises, is whether design does useful work (2002, 354). Exactly, and until there is an honest attempt to face up to evidential issues, the design hypothesis is nothing more than a wheel spinning in a Rube Goldberg machine that fritters away useful energy, contributing to the entropic disorder of the world, while doing no discernible explanatory work of value.

Dembski on Design

Earlier in this chapter, we saw that Dembski's complexity-specification criterion singles out irreducibly complex biochemical systems as intelligently designed systems. Since we are questioning the conclusion that such biochemical systems are in fact the fruits of intelligent design, it will not go amiss here to examine Dembski's criterion in more detail. Dembski considers messages from space as a means of illustrating his complexity-specification criterion. The specification part of the criterion is there to ensure "that the object exhibits the type of pattern characteristic of intelligence" (2001b, 178). He rightly observes that the SETI (Search for Extraterrestrial Intelligence) program involves scientists actively seeking to detect messages from space. Such scientists need to take steps to filter out signals of interest from those that can be produced by background noise or natural mechanisms such as pulsars.

A *signal of interest* here is a nonrandom sequence of events that merits further analysis. It is probably brought about by some mechanism or other, *possibly* an alien intelligence. What would we look for in such a signal in order to infer design? Following St. Thomas Aquinas (discussed in chapter 1), Dembski directs our attention to lessons derivable from archery. Suppose I am alone and shooting arrows at the side of a large barn. Suppose I wish to impress my friends with my skill at archery. I shoot from a distance of fifty feet. Having hit the wall of the barn with all my arrows (even I can't miss the side of a barn at fifty feet), I then go and paint bull's-eyes around them and call my friends over to have a look. What can my friends conclude when they arrive at the barn? Dembski tells us:

> Absolutely nothing about the archer's *ability as an archer*. Yes a pattern is being matched, but it is a pattern fixed only after the arrow has been shot. The pattern is thus purely *ad hoc*.
>
> But suppose instead the archer paints a fixed target on the wall and then shoots at it. Suppose the archer shoots a hundred arrows, and each time hits a perfect bull's-eye. What can be concluded from this second scenario? *Confronted with this second scenario we are obligated to infer that here is a world class archer, one whose shots cannot legitimately be referred to luck, but must rather be referred to the archer's skill and mastery. Skill and mastery are of course instances of design*. (2001b, 180, my italics)

This is all well and good *if we see the archer do the trick*. The trouble here should be obvious. We have to ask what happens if we do not see the archer shoot and we do not see the arrows in flight. We have to ask how we would react to being presented simply with a target whose bull's-eye was chock full of arrows. My friends, who were not impressed with the first case described, would be even less impressed with my ability as an archer if I simply confronted them with a straw target whose bull's-eye was chock full of arrows and asked them what they thought of my skill and mastery as an archer.

They would no doubt say that they could not see evidence of skill and mastery at archery. My friends would want to see me hit the bull's-eye several times from fifty feet; they would want appropriate analogs of the intelligent design questions answered through the provision of high-quality evidence of intelligent, skillful archery. The bull's-eye full of arrows would not, in and of itself, be enough. This is why the origins of pattern, not simply the pattern itself, are important. Without this, all you have is anomalous data. Behe is simply wrong when he says, "The inference to design can be held with all the firmness that is possible in this world, *without knowing anything about the designer*" (1996, 107, my italics).

What are we to conclude about radio signals from space? How, in addition to contingency and complexity, would we be able to recognize specificity? In the movies, the scientists receiving the signal run it through a computer, and a translation is quickly produced. Perhaps the message is targeted at us. For example, "We don't love Lucy! More *Gilligan's Island* please." Perhaps it is a general call. For example, "Hi, is anybody there?" Perhaps we have eavesdropped on communications not intended for us. For example, "Zort loves Snurt." But this is science fiction. How could we tell from a radio signal, in and of itself, that it contained complex specified information indicating intelligent origins?

Neither of Dembski's attempts to answer this question is satisfactory. In the first case, he draws an analogy with human speech and human writing (both of which are known antecedently to be phenomena that, under a wide range of conditions, can exhibit varying degrees of intelligence and design). He says, "Whenever a human being utters meaningful speech, a choice is made from a range of possible sound-combinations that might have been uttered. Intelligent agency always entails discrimination, choosing certain things and ruling out others" (1999, 144). Alternatively, what is the difference between ink spilled on a sheet of paper and a written message? Dembski tells us, "A random ink blot is unspecified; a message written with ink on paper is specified. To be sure, the exact message recorded may not be specified. But orthographic, syntactic and semantic constraints will nonetheless specify it" (1999, 145).

Once again, signals from space are not like marks on paper, where there is antecedently recognizable handwriting (orthography), grammar (syntax), and meaning (semantics). Of course, if you have all this, it's a no-brainer. But if not, what you have are nonrandom marks on paper, brought about by a mechanism that may or may not have been intelligent. I have such a letter from a friend's young child. As Elsberry has noted:

> SETI can only detect signals that possess certain properties known from prior experience of humans communicating via radio wavelengths. SETI works to find events that conform to our prior experience of how intelligent agents use radio wavelengths to communicate. SETI does not support the notion that novel design/designer relationships can be detected. ETI that communicate in ways outside human experience will be invisible to, and undetected by, SETI. (1999, 34)

What we need is an example that does not involve humans or imaginary aliens from outer space. And Dembski provides just that.

The second case involves an appeal to scientific studies of animal behavior. Of rats negotiating mazes, Dembski writes:

> Only if the rat executes the sequence of right and left turns specified by the psychologist will the psychologist recognize that the rat has

> learned how to traverse the maze. Now it is precisely these learned behaviors that we regard as intelligent in animals. Hence it is no surprise that the same scheme for recognizing animal learning recurs for recognizing intelligent agency generally: actualizing one among several competing possibilities, ruling out others and specifying the one actualized. (1999, 145)

Here I can only assume that Dembski's lack of acquaintance with relevant literature on animal behavior has let him down. These issues are reviewed in a readable and entertaining manner by Marian Dawkins (1998), and Sara Shettleworth (1998) and Euan Macphail (1998) provide more technical discussions.

One of the central challenges that face those who study animal behavior is that of deciding exactly what sort of animal behavior signals intelligence and cognitive sophistication. The issue is far from trivial and can concern behaviors that are sometimes much more complicated than maze negotiation, such as the behaviors we find in studies of deception and self-consciousness in primates. One of the central questions that arises in the experimental context is whether behavioral data can be explained in terms of associative learning (stimulus-response effects), rules of thumb (sequences of simple behavioral rules whose collective effect is complex, seemingly intelligent behavior), or genuine cognitive sophistication.

Behavioral data can be horribly misleading, as anyone acquainted with the "Clever Hans" effect knows only too well. Clever Hans was a horse that could perform amazing feats. But the cleverness turned out to be learned responses to subtle behavioral cues from his handler (see Dawkins 1998, 68–71). The same effect undermined early studies on language use in chimpanzees. Contingent, complex, nonrandom, specified behavioral patterns do not necessarily herald the existence of intelligence.

So to return to Dembski's discussion of animal intelligence, he tells us that if we were to observe a rat exit a sophisticated maze quickly and efficiently, we would be convinced that "the rat has indeed learned to exit the maze and that this was not dumb luck" (1999, 146). Agreed, but this particular application of the complexity-specification criterion is problematic. The rat's complex, maze-negotiation behavior is surely evidence that blind luck and purely random behavior is not the explanation. Here I agree with Dembski. But it does not follow that we have evidence of intelligent agency as opposed to a lower level explanation. In the study of animal behavior, you cannot infer from the fact that a behavior is not due to luck that it results instead from intelligent causes (as opposed to mindless conditioning or the use of rules of thumb—simple behavioral rules, resulting perhaps from genetic predispositions that generate the appearance of intelligence).

Here is an example borrowed from Marian Dawkins's discussion of these issues. A bird visits the territories of several males and selects one of them as a mate. It might be that she is visiting each male, consciously assessing him for genetic worth, and comparing him with others she has observed. Television shows about animals routinely describe their amazing behaviors in similar terms. But it might be that she is simply disposed (perhaps by inheritance) to mate with the male with the longest tail or the loudest call. As Dawkins herself observes:

> Faced with the complexity of animal behavior (and animals are genuinely far more complex than any man-made machine so far devised), we have a tendency to jump to the conclusion that it is much more complicated and mysterious than it really is. Because we don't understand fully how animal bodies function, we tend to assume that they achieve their complexity by thinking and working things out. But before we are entitled to conclude that that is really what they are doing . . . we must be sure that what we are looking at could not be explained much more simply with a rule of thumb. A switch in a hormone level or a greater response to a long tail than a short one is a much simpler explanation of why a female mates with one male rather than another than would be implied by saying that she "assesses every male in turn." These rules of thumb can be very difficult to spot and very deceptive in leading us to think that something is complex when it really turns out not to be after all. (1998, 86–87)

Careful, controlled experimental studies that go deeply into an analysis of mechanisms and go well beyond straightforward examinations of behavioral patterns are needed to deal with these issues.

Another way to bring this worry to the fore is to observe that Dembski's design inference is an inference from a presented pattern of events. As he himself notes (2002, 165), intelligent agents can mimic patterns that normally arise from natural causes. But as the case of animal behavior shows, it is possible for unintelligent animals to generate behavioral patterns that mimic those that might be produced by intelligent agency. Moreover, you cannot tell simply by looking at, say, maze negotiation patterns, which mechanism—intelligence, conditioning, or rules of thumb—is responsible. Yet the design inference depends solely on the presented pattern of events, not on an analysis of what caused the pattern, and is for this reason apt to be highly misleading.

In addition, as with the SETI case discussed by Elsberry, a related worry is lurking for the incautious student of animal behavior. I have recently shown (2002) that what constitutes evidence of intelligent behavior in animals is not a simple issue to be settled by examining raw data. Different theoretical perspectives (for example, the very big theoretical differences between behaviorists, who prefer explanations of animal behavior in terms of associative learning, and cognitive ethologists, who want to explain animal behavior by using rich cognitive attributions of consciousness, beliefs, hopes, fears, desires, and so on) give very different interpretations of the same experimental data and result in very different sorts of inference from the data. Once again, the raw, unanalyzed pattern tells you very little in and of itself.

However these issues are to be settled, Dembski reminds us, "*Ultimately what enables irreducible complexity to signal design is that it is a special case of specified complexity*" (2002, 115, my italics). With this in mind, we must return to biochemistry. Perhaps here at last we have something that could not be explained, even in principle, using natural, unintelligent mechanisms.

Self-Organization as a Route to Irreducible Complexity

Following the discussion in chapter 3, in which we saw how the Second Law of Thermodynamics permitted the emergence of complex, ordered, organized states of matter through the mechanism of self-organization, the question naturally arises as to whether chemical self-organization could give rise to an irreducibly complex system.

In our work on biochemical complexity, Karl Joplin and I suggested that you could get an irreducibly complex system to self-organize in the laboratory. In essence, we claimed you could get a chemical equivalent of an irreducibly complex pocket watch to self-organize in a beaker—hence the case was relevant to Paley's argument as well as Behe's. The reaction we looked at was the famous Belousov-Zhabotinski (BZ) reaction (Shanks 2001b; Shanks and Joplin 1999, 2001a). Since both Behe and Dembski have made comments about this example, it will not go amiss to examine the case further.

The BZ reaction refers to a set of chemical reactions in which an organic substrate is oxidized in the presence of acid by bromate ions in the presence of a transition metal ion (Tyson 1994, 569). The version of the reaction that Karl Joplin and I use in our classroom has the following ingredients: potassium bromate, malonic acid, potassium bromide, cerium ammonium nitrate, and sulfuric acid. When ingredients are placed in a beaker, the system self-organizes to perform a repeating cycle of reactions. It behaves as a chemical oscillator, and the oscillations can be monitored through cycles of color changes. You can use it to tell the time! It is, in essence, a Breguet repeater in a beaker (so much more complicated than a mere mousetrap).

The oscillations result from the chemical system cycling through its component reaction pathways. What does this mean? Well, suppose the system starts out with a high concentration of bromide ions. In the first group of reactions, bromate and malonic acid are used in a slow reaction to produce bromomalonic acid and water. Bromous acid is one of the reaction intermediates in this pathway. Since the cerium present is in the cerous state, the reaction medium remains in the colorless state for this phase of the cycle. As time goes by, the concentration of bromide ions drops to a point where bromous acid can initiate another mechanism to produce bromomalonic acid and water.

Here, in a fast reaction, bromate, malonic acid, bromous acid (a reaction intermediate from the first pathway), and cerous ions produce ceric ions, bromomalonic acid, and water. The reaction medium turns yellow as cerium enters the ceric state. The pathway also contains an autocatalytic step, like the one in the rabbit-grass pathway discussed in chapter 3, in which the presence of bromous acid catalyzes the production of more of itself, so one mole of bromous acid makes two moles of bromous acid. (This positive feedback effect is why this pathway is fast.) As cerous ions are consumed and ceric ions accumulate, a critical threshold is achieved, at which time a third pathway opens up. This pathway consumes bromomalonic acid produced by the previous two pathways, malonic acid, and ceric ions to produce carbon dioxide and bromide ions and to regenerate cerous ions, thereby setting the system up for a new cycle (see Babloyantz 1986, 158–159 for details).

The Second Law is not violated. To get the oscillations, the system begins far from chemical equilibrium. The oscillations continue until equilibrium is reached (the period of oscillation gradually getting longer and the color changes less pronounced as equilibrium is approached). Karl Joplin and I have had the system oscillate for more than an hour in typical classroom demonstrations, much to the distraction of our students. That the reaction manifests self-organization means nothing more than that the invisible hand of the chemical interactions between molecules, in accord with the laws of chemistry, brings about highly ordered, coherent behavior of the system as a whole in the form of regular temporal oscillations. The explanation of this behavior does not require the intervention of a supernatural intelligence.

Joplin and I argued that the BZ system manifested irreducible complexity because it satisfies all the requirements of the mousetrap model of irreducible complexity. Behe tells us that there are three steps to be satisfied. The system must have a function. Behe tells us, "The function of the system is determined from the system's internal logic" (1996, 196). In the light of this, the function of the BZ reaction—determined by the logic of the chemical interaction dynamics internal to the system—is to oscillate.

The next requirement is that the system must consist of several components (1996, 42). The BZ system consists of several key reactions. Behe does not appear to dispute this part of our example. Finally, we must ask whether all the components so identified are required for the achievement of function. The key components of the BZ reaction are all needed for the oscillatory cycle to exist. The disruption of any of these key reactions results in the catastrophic failure of the system. Apparently, the dumb, unguided laws of chemistry will generate irreducibly complex systems.

Yet Behe has objected to our example. It is instructive to examine his reasoning. Commenting on the BZ system, he notes, "Although it does have interacting parts that are required for the reaction, the system lacks the crucial feature—the components are not *well-matched*"

(2000, 157, my italics). Behe tells us that only systems that require well-matched components are irreducibly complex. Behe then objects to the example of the BZ reaction by arguing that the reagents used in the BZ reaction have a wide variety of uses—in Behe's terminology, they have low *specificity*. To be well matched, as Dembski has observed in his discussion of Behe's argument, is to be like the fan belt of a car, "specifically adapted to the cooling fan" (2002, 283).

In this light, Behe argues that one ingredient of the BZ system, sodium bromate, is a general-purpose oxidizing agent, and ingredients other than the ones we mentioned can be substituted. In our reaction, we mentioned the use of cerium ions, but other types of ions will work just as well. Behe points out that the reaction is easy to set up and runs over a wide range of concentrations (2000, 158).

Fair enough. But if Behe is right about all this, then mousetraps are not irreducibly complex either. Their components also have low specificity. The steel used in their construction has a wide range of uses, as does the wood used for the base. You can substitute plastic for wood, and any number of metals for the spring and hammer. Mousetraps are easy to make (which is why they are cheap) and will work with metals manifesting a wide range of tensile strengths. Either the BZ system is an irreducibly complex system, or the complexity of mousetraps is not a model for irreducible complexity. Take your pick, for you cannot have it both ways.

This matter is made all the more acute, however, because crucial components of Behe's own biochemical examples of irreducible complexity have multiple uses and lack substrate specificity (interact with a wide variety of substrates). For example, plasminogen (a component of the irreducibly complex blood-clotting cascade) has been documented to play a role in a wide variety of physiological processes, including tissue remodeling, cell migration, embryonic development, and angiogenesis, as well as wound healing (Bugge et al. 1996).

And though Behe tells us that plasmin (the activated form of plasminogen) "acts as scissors specifically to cut up fibrin clots" (1996, 88), we learn in one of the very papers he cites, "*Plasmin has a relatively low substrate specificity and is known to degrade several common extracellular-matrix glycoproteins in vitro*" (Bugge et al. 1996, 709, my italics). Plasmin, it would seem, does not appear to be well matched like the fan belt of a car, "specifically adapted to the cooling fan." Behe cited this paper, so he knows this as well as I do.

Redundant Pathway Complexity

In our initial analysis of Behe's work, Karl Joplin and I proposed that many real biochemical systems exhibited redundant complexity rather than irreducible complexity. Behe now concedes that biochemical systems can indeed exhibit redundancy (2000, 160). This is no mean concession, since biochemical redundancy plays a crucial role in the explanation of why biological systems are evolvable (See Gerhart and Kirschner 1997, 588–589). So what is redundant biochemical complexity? And why should we care about it?

We see redundant complexity when we notice that many actual biochemical processes do not involve simple linear sequences of reactions, with function destroyed by the absence of a given component in the sequence. Instead, they are the product of a large number of overlapping, slightly different—hence redundant—processes. Redundant complexity is also embodied in the existence of backup systems that can take over if a primary system fails. Finally, redundant complexity is observed in the phenomenon of convergent biochemical evolution, whereby systems with different evolutionary histories, perhaps using different substrates and products, nevertheless achieve similar biochemical functions.

The redundant complexity of biochemical processes turns out to lie at the heart of the stability, flexibility, and robustness they manifest in the face of perturbations that ought to catastrophically disrupt systems conceptualized from the standpoint of Behe's metaphor of the well-designed, minimalist mousetrap—the absence of any component of which should render the system functionless. To better understand redundant complexity, it will help to look at some examples.

If we examine the central catabolic pathway of glycolysis (the interconnected series of reactions by which glucose is broken down to release usable energy), superficially it looks like the product of one reaction in the series is required as the substrate for the next reaction in the sequence:

$$\text{[1]} \qquad A - a \rightarrow B - b \rightarrow C - c \rightarrow D \ldots,$$

where A is the initial substrate (glucose), *B* is the product (glucose-6-phosphate), and *a* is the enzyme (hexokinase) catalyzing (or promoting) the transformation of *A* into *B*. *B*, the product of the first step in the pathway, then becomes the substrate of the second step to be turned into product C, with the help of enzyme catalyst *b*, and so on.

Thinking with the aid of a "glycolysis mousetrap model," one might expect that removing one component, either enzyme, substrate, or product, would shut down the pathway and prevent the continual production of energy. In fact, almost every step in this pathway is redundantly complex. Lets focus on the first step: the production of glucose-6-phosphate from glucose, catalyzed (promoted) by the enzyme hexokinase.

Not only does hexokinase activate the relatively stable glucose (Bennett and Steitz 1978) but also it is a multipurpose enzyme that, in part, controls the rate of the first part of the glycolytic pathway by directing the chemistry of glucose to either build up more complex molecules (anabolism) or harvest the energy stored in glucose (catabolism). The direction of chemical activity is dependent only on the concentration of the substrates, products, and various components of the pathway (Voet and Voet 1995).

One might assume, therefore, that we have a good example of Behe's mousetrap. Remove the enzyme, and the reaction should stop. But this intuition rests only on a superficial characterization of this step in the pathway. Looking at the fine details reveals an unexpected complexity to what appears to be a simple, straightforward chemical situation. In typical vertebrate tissue, redundant complexity is manifested in the existence of several different variants (isoforms) of hexokinase. All are present, as a result of gene duplication and differential expression, in varying proportions, in different tissues. Removal of a given variant of hexokinase does not disrupt glycolysis, though it may have an effect on the efficiency with which a function is achieved. Depending on whether the tissue requires rapid utilization of energy (muscles) or is involved in converting the glucose to a storage form, glycogen (in the liver), the proportions of the variants differ for these specialized functions. So there is redundant complexity here, in the first step of the glycolytic pathway, a seemingly simple, straightforward step.

Each of the components of the rest of the glycolysis pathway manifests similar redundancies. Remove glucose, and the pathway can utilize numerous other hexose (six-carbon atom) sugars to supply the next product. Knock out an enzyme, and the "glycolysis mousetrap" should fail. But which enzyme? Knock out one enzyme variant, and the other variants in the tissue can take over its function. Maybe not quite as efficiently, but, as Behe concedes, efficiency is something that can be improved by natural selection over evolutionary time. There are backup systems, too. For example, if you could succeed in removing all the variants of hexokinase, there are alternative pathways that can supply the needed products, such as the pentose phosphate pathway (Martini and Ursini 1996).

It is a hallmark characteristic of many evolved biochemical systems that there are typically multiple causal routes to a given functional end, and where one route fails, another can take over. The existence of variants of a given enzyme are evolutionary legacies—legacies by means of which one and the same enzyme can be co-opted to serve several specialized functions in specialized tissues. Karl Joplin and I (1999) discussed many different examples of redundancy in biochemistry and molecular biology.

Redundant Complexity (Again)

Behe concedes the existence of redundant biochemical complexity. His discussion of the way Joplin and I used the concept takes us to the heart of the issue of the intelligent design of biochemical systems. As Behe rightly observes, "The observation that some biochemical systems are redundant does not entail that all are. And in fact, some are not redundant" (2000, 160). Behe goes on to give some interesting examples to make his case. Let's suppose he is right. This raises the crucial question of the origins of this irreducible complexity.

There are at least two hypotheses concerning the origins of irreducibly complex biochemical systems:

> (H1) Irreducibly complex biochemical systems arose as the result of supernatural intelligent design. (We know not how, by whom, when, or for what purpose.)

With this hypothesis in mind, Behe has observed: "As an important corollary, it also predicts that mindless processes such as natural selection or the self-organization scenarios favored by Shanks and Joplin will not be demonstrated to be able to produce irreducible systems of the complexity found in cells" (2001a, 15). Behe nowhere demonstrates this important corollary, and without further argumentation, it cannot safely be concluded (since Behe and his friends refuse to say anything illuminating about the designer) that the hypothetical designer did not arrange for things to occur by the mindless mechanisms favored by Karl Joplin and my good self.

In view of this, another hypothesis to consider is as follows:

> (H2) The mindless process of Darwinian natural selection can produce irreducibly complex systems of the kind found in cells.

Since H1 says nothing more than that the origins of irreducible complexity lie in opaque supernatural causes, we had better focus our attention on H2.

In terms of redundant complexity, we have the tools to provide a naturalistic, evolutionary explanation of the source of irreducible complexity. Behe's metaphor for an irreducibly complex system was that of the mousetrap. Karl Joplin and I have argued a better metaphor is found in the architectural image of a free-standing arch. This image was first suggested by A. G. Cairns-Smith, a biochemist whose writings about the origins of biochemical complexity predate Behe's work by a decade or more (1986, 59–60).

Cairns-Smith's own interests were in the origins of life, but the complexity problem he confronted was essentially identical to that raised a decade later by Behe—though couched in different terminology. It is wrong, therefore, to suppose that Behe discovered a special complexity problem that was hitherto unknown in the biochemical community. For these reasons alone, it is instructive to examine Cairns-Smith's reasoning a bit more closely.

Cairns-Smith's complexity problem was discussed under the heading of the unity of biochemistry, but Behe's problem, stated a decade later, is clearly very similar. Thus Cairns-Smith comments:

> For example, proteins are needed to make catalysts, yet catalysts are needed to make proteins. Nucleic acids are needed to make proteins, yet proteins are needed to make nucleic acids. Proteins and lipids are needed to make membranes, yet membranes are needed to provide protection for all the chemical processes going on in the cell. . . . *The interlocking is tight and critical. At the center everything depends on everything.* (1986, 39, my italics)

Cairns-Smith thinks this complexity must be explained. However, unlike Behe, Cairns-Smith thinks a natural, rather than a supernatural, explanation will suffice.

It is at this point that we are invited to consider a free-standing arch of stones. It manifests irreducible complexity in that the keystone at the top of the arch is supported by all the other stones in the arch, yet these stones themselves cannot stand without the keystone. In other words, the arch stands because all the component stones depend on each other. Take away a stone, and the arch collapses. However, Cairns-Smith notes, not all the stones, nor all the functional biological structures, must be there from the beginning:

> It is clear that not all such functions were hit on at once. Some would have been later discoveries. If new uses may be found for old structures, so, too, can old needs be met by more recently evolved structures. There is plenty of scope for the accidental discovery of new ways of doing things that depend on two or more structures that are already there. . . . This is typical at all levels of organization, from organs to molecules. (1986, 59)

He adds: "There is plenty of scope for accidental discoveries of effective new combinations of subsystems. It seems inevitable that every so often an older way of doing things will be displaced by a newer way that depends on a new set of subsystems. It is then that seemingly paradoxical collaborations may come about" (1986, 59). Why does he think these collaborations are paradoxical?

Referring back to the stone arch, Cairns-Smith anticipates Behe by observing, "*This might seem to be a paradoxical structure if you had been told that it arose from a succession of small modifications, that it had been built one stone at a time*" (1986, 59, my italics). This is especially true if, as in biochemistry, the arch is multidimensional, with central "stones" each touching more than the two stones touched by the keystone in our arch (1986, 60).

Nevertheless, it is possible to construct an arch in gradual stages. You cannot, of course, gradually build a self-supporting, free-standing arch by using only the component stones, piling them up, one at a time. But if you have scaffolding—and a pile of rocks will suffice to support the growing structure—you can build the arch one stone at a time until the keystone is in place, and the structure becomes self-supporting. When this occurs, the (now redundant) scaffolding can be removed to leave the irreducibly complex, free-standing structure. In this way, the redundant complexity of biochemical systems, whose existence Behe concedes, can be employed to explain the origins of irreducibly complex systems.

Natural, mindless evolutionary processes give rise to the redundant complexity we observe in biochemical systems. These redundancies then provide, in concert with extant functional systems and structures, the biochemical and molecular scaffolding to support the gradual evolution of systems that ultimately manifest irreducible complexity when the scaffolding is removed or reduced. The resulting biochemical arches may then achieve functions as integrated wholes that could not be achieved by the parts acting independently. Natural selection will result in some of these biochemical arches being retained for further evolutionary elaboration, while others will be eliminated by the same mechanism. Irreducibly complex systems can simply be viewed as limiting cases of redundantly complex systems. Reduce redundancy to the point where further reduction results in loss of function, and the system is now irreducibly complex.

Irreducible complexity was supposed to be something that could not, even in principle, be explained by Darwinian methods. It is now clear that it can indeed be explained in principle by using Darwinian methods. And there matters would rest, had it not been for the recent intervention of Dembski, who has deftly tried to move the target that Darwinians are supposed to hit with their scientific arrows:

> The scaffolding objection has yet to demonstrate causal specificity when applied to actual irreducibly complex biochemical systems. The absence of detailed models in the biological literature that employ scaffoldings to generate irreducibly complex biochemical systems is therefore reason to be skeptical of such models. *If they were the answer, then one would expect to see them in the relevant literature, or to run across them in the laboratory. But we do not. That, Behe argues, is good reason to think they are not the answer.* (2002, 254, my italics)

It is perhaps ungentlemanly of me to point out that this demand for causal specificity and concrete experimental details is coming from one who advocates supernatural design, about which apparently nothing whatsoever can be said. But ungentlemanly or not, I believe Dembski is mistaken.

We will approach the matter by degrees, gradually building up to a consideration of some of Behe's specific examples. The study of developmental processes suggests that an important biological role is indeed played by removable scaffolding, in the formation of all manner of elaborate structures, including body parts and neural pathways. For example, developmental scaffolding, in the form of an initial superabundance of cells, can be removed by apoptosis, the biochemical activation of self-destruction genes, and this process plays a crucial role in the developmental sculpting of such structures as fingers and toes (Campbell 1996, 980; Lewis 1995, 15).

Developmental biologists are very interested in characterizing developmental pathways by which an organism gets from one developmental state to another. In such a pathway, we might have a gene-mediated step as follows:

[2] $$A - a \rightarrow B,$$

where A and B are successive developmental states and a is the gene whose biochemical product mediates the step. But we can also have redundancy:

[3] $$A - (a + b) \rightarrow B.$$

Of this case, Wilkins has recently observed:

> If, however, two gene products contribute to the same step, and their activities are similar and additive at this step, then mutational inactivation of one gene will often be masked by the continued activity of the other.... The consequence is that mutational inactivation of either gene is frequently insufficient to block the sequence, and correspondingly, activity of both genes must be eliminated to prevent step *B* from occurring. *In general, pathway steps with dual, or multiple, inputs of this kind will be missed in conventional mutant hunts, since, in general, only a single gene of the pathway is affected in each mutant line.* (2002, 114)

So not only do we learn here that redundant scaffolding is important but also we learn that scientists in the laboratory do look for disruption of function in otherwise irreducibly complex systems by performing mutant hunts on carefully inbred strains of research animals. These hunts are complicated by the existence of redundancy. Wilkins (2002, 114–116) provides some examples. The point is that irreducibly complex developmental pathways can be viewed simply as limiting cases of redundantly complex pathways. Reduce redundancy to the point where further reduction results in loss of function, and the pathway is now irreducibly complex.

As noted previously, gene duplication is one route to redundant complexity, but how could redundancy be reduced to give rise to irreducible complexity? One way is through the transformation of functional genes into *pseudogenes*—nonfunctional members of gene families. Molecular biologist H.-S. Li observes, "Pseudogenes are DNA sequences that were derived from functional genes but have been rendered nonfunctional by mutations that prevent their proper expression. Since they are subject to no functional constraints, they are expected to evolve at a high rate" (1997, 187).

If a functional gene becomes a pseudogene, its product will no longer be available to the biochemical pathways in which it formerly participated. The transformation of a gene to a pseudogene will not have catastrophic consequences if the biochemical pathways in which its product formerly participated are redundantly complex—other products can take over the role of the missing product. Perhaps not as efficiently, but efficiency is something that can be improved by selection. In this way, redundant scaffolding can be reduced, ultimately to the point where a system or pathway is irreducibly complex. There may be strong selection against further reductions at this point of evolution, but not necessarily, as we shall see later.

Behe's Examples Revisited

Now that we have an evolutionary framework within which we can explain the origins of irreducibly complex systems, we can usefully reexamine Behe's examples.

Behe cites as an example of an irreducibly complex system the synthetic pathway that makes vitamin C in other mammals but which in humans (and certain other animal species) is disrupted by the lack of a functional gene for L-gulono-gamma-lactone oxidase. But as he himself observes, in humans a pseudogene is present (2000, 160). The vitamin C "mousetrap" ceases to function in humans when a functional component is lost. But this is hardly a shocking observation. The pathway in humans has simply been disrupted by a continuation of the same sorts of processes that reduced redundancy to yield irreducibly complex systems in the first place. Seen in proper Darwinian light, Behe himself answers Dembski's challenge!

From an evolutionary standpoint, this example looks like a case of use it or lose it. As Nesse and Williams comment: "Our ancestral shift to a high-fruit diet, rich in vitamin C, had the incidental consequence about forty million years ago of allowing the degeneration of the biochemical machinery for making this vitamin. Our frugivorous close relatives share our requirement for dietary vitamin C" (1994, 130). In this case, loss of a functional pathway was preceded by adaptation to a niche rich in vitamin C.

Mutational events called *deletions*—whereby bases are deleted from genes (often a single base or a few bases, but sometimes several thousand)—occur naturally and can result in dysfunction. A special kind of deletion, however, is artificially induced in a knockout experiment. In a knockout experiment, a gene is deliberately deleted from a genome, and hence all the causal roles played by that gene are halted (Travis 1992).

Researchers can now target a specific gene in mice and knock it out. Such knockout mice are valuable models for human diseases in gene function experiments. However, such mice do not always give the expected result—they do not exhibit the predicted functional deficits—and when this happens, they serve as examples of the type of redundant complexity we have been discussing. Thus, Gerhart and Kirschner note: "*Gene knockout in mice, where a specific gene is deleted or rendered inactive, is the crucial test of genetic redundancy in metazoans*. Gene knockouts do, of course, represent a nonrandom selection of mutations, since the genes that have been disrupted were chosen for study because they were expected to be critical for function" (1997, 589, my italics). Apparently the literature is not as devoid of experiments to examine redundancy (scaffolding) and irreducible complexity after all.

One example concerns the gene *p53*, originally identified as a tumor suppression gene, which has subsequently been found to be involved in a number of fundamental cell processes. For example, it plays a role in gene transcription, the cell cycle, programmed cell death (apoptosis), DNA replication, and DNA repair processes (Elledge and Lee 1995).

Looking at this case from the standpoint of a genetic mousetrap model, one would naturally predict that the removal of this gene, involved as it is in all of these vital processes, would lead to catastrophic collapse of the developmental process—a bit like removing the spring, trigger, or platform from Behe's mousetrap. Such is not the case, since *p53* knockouts in mice yield viable, fertile offspring, although they are susceptible to the early appearance of spontaneous tumors (Dowehower et al. 1992). This suggests the following dilemma: Either *p53* is not required for embryonic development, or there are redundant ways in which the function of the missing component is compensated for (Elledge and Lee, 1995). The evidence at hand supports redundant complexity, since there are at least 400 proteins associated with the proper control of the cell cycle alone (Murray and Hunt, 1993), and it would appear that some of these other proteins pick up the slack created by the missing *p53*. Such mice can still be caught in mousetraps!

Behe acknowledges this case but draws his reader's attention to the blood-clotting cascade originally discussed in his book: "Yet contrast this case [*p53*] with that of mice in which the gene for either fibrinogen, tissue factor, or prothrombin has been knocked out. . . . The loss of any one of those proteins prevents clot formation and the clotting cascade is broken. Thus Shanks and Joplin's concept of redundant complexity does not apply to all biochemical systems" (2000, 161). Exactly right. Loss of functional genes reduces redundancy to yield an irreducibly complex system. All Behe's example shows is that further losses at this point can catastrophically disrupt the system.

It is possible, however, that Behe and Dembski have both made too much of the examples that Behe has employed in his discussions of irreducible complexity. It is worth noting, for example, that the blood-clotting cascade itself has features that manifest redundant complexity. The real situation is thus more complex than Behe's carefully massaged description would lead you to believe. Plasminogen deficient (*Plg-/-*)—hence plasmin deficient—mice have been studied. As noted earlier, plasmin is needed for clot degradation (it cuts up the fibrin), yet as Bugge et al. comment:

> Plasmin is probably one member of a team of carefully regulated and specialized matrix-degrading enzymes, including serine-, metallo-, and other classes of proteases, which together serve in matrix remodeling and cellular reorganization of wound fields. . . . *However, despite slow progress in wound repair, wounds in Plg-/- mice eventually resolve with an*

> *outcome that is generally comparable to that of control mice. Thus an interesting and unresolved question is what protease(s) contributes to fibrin clearance in the absence of Plg?* (1996, 717, my italics)

Once again, Behe cited this very paper, so we must assume that he, too, knows that parts of his clotting-cascade are redundantly complex.

Conclusion

This chapter has involved a long journey into strange territory. The end result, however, is that, notwithstanding repeated claims to the contrary, we have not been given the extraordinary evidence we need to rationally accept the extraordinary claim that biochemical systems are the fruits of supernatural intelligent design, or any other kind, for that matter. But if we organisms are not the fruits of intelligent design, does this mean there is no God? Kenneth Miller, in his book *Finding Darwin's God*, has observed:

> We know from astronomy that the universe had a beginning, from physics that the future is both open and unpredictable, from geology and paleontology that the whole of life has been a process of change and transformation. From biology we know that our tissues are not impenetrable reservoirs of vital magic, but a stunning matrix of complex wonders, ultimately explicable in terms of biochemistry and molecular biology. With such knowledge we can see, perhaps for the first time, why a creator would have allowed our species to be fashioned by the process of evolution. (1999, 290)

If Miller is right, you can reject antievolutionary intelligent design, accept evolutionary explanations, and still be content in your belief in God. So could there be some other source of extraordinary evidence we have overlooked—evidence that will bear some theological weight?

To examine this issue we need to examine what science has uncovered about the origins of our universe. In the next chapter, we will examine claims about design that proceed from considerations about the nature of the universe. Can the cosmological arguments for intelligent design do better than the biological arguments we have just examined?

6

The Cosmological Case for Intelligent Design

In the tales told by ancient Middle Eastern sheepherders, in the beginning, God made heaven and earth. According to Archbishop James Ussher (1581–1656), who had carefully worked his way through the records of these tales, the creation took place in 4004 B.C., on October 23 at 12:00 P.M. This conflicts with modern science, according to which the planet Earth formed about 4.5 billion years ago in a universe that itself originated in the big bang event some 14 billion years ago.

Biological versions of the argument from design, as we have just seen, hold no water. There are no secret signatures of the designer in our molecules or the biochemical pathways in which they function. But perhaps we have looked for the evidence of design in the wrong place. Perhaps it is not to be found in our molecules or our organs, such as our eyes. Perhaps it is to be found in the universe at large and in the way the universe reflects important properties of its tiniest microscopic—indeed, subatomic—constituents.

This at least is the hope of those who point to what have become known as the *anthropic coincidences* uncovered by contemporary astrophysics and cosmology. We will shortly examine these coincidences and how intelligent design theorists interpret them. For the present, I note that they are viewed by design theorists as further examples of complex-specified information, and hence as evidence of intelligent design (Dembski 1999, 159). That scientists have uncovered features of the universe known as anthropic coincidences cannot be denied, but as you might have guessed, the trouble lies not in the phenomena, in and of themselves, but in the explanations and interpretations that are offered to render them meaningful.

Before we can examine the anthropic coincidences that some theorists see as evidence of design, we need to know a bit more about the way in which modern science treats the early universe and the events surrounding its origins in the cosmic event that has come to be known as the big bang. Kenneth Miller, in his book *Finding Darwin's God*, observes:

> One of the most remarkable findings of cosmological science is that the universe did have a beginning, and a spectacular beginning at that. Discussions of first causes used to be dry philosophical constructs, theoretical arguments against an infinite regression of events backwards in time. The big bang made the first cause real. It placed a wall at the beginning of time, closing to inquiry (but not, of course, to speculation) all events that might have occurred before the cosmic explosion. In the view of many scientists, the big bang casts a distinctly theological light on the origin of the universe. (1999, 225)

Is cosmology, then, a place where science can have a meaningful dialogue with religion? Perhaps even a dialogue in which religion can embrace science instead of desperately denying it? I do not think so, and I will now try to explain why intelligent design theory fares no better in cosmology than it does in the context of biology or biochemistry. Readers interested in learning more about cosmology might consider looking at Weinberg (1988), Hawking (1988), Hogan (1998), or Rees (2001). Gale and Shanks (1997) discuss the historical background to the rise of modern cosmology. The anthropic coincidences are examined in a readable way in Gale (1981, 1990, 1997) and Rees (1999).

The Big Bang

Why did cosmologists come to the idea that the universe had a beginning? Why not suppose it had simply existed forever? That there must have been a beginning to our universe was an idea that was suggested by astronomical data. Stephen Hawking observes:

> In 1929, Edwin Hubble made the landmark observation that wherever you look, distant galaxies are moving rapidly away from us. In other words, the universe is expanding. This means that at earlier times objects would have been closer together. In fact it seemed that there was a time, about ten or twenty thousand million years ago, when they were all exactly at the same place and when, therefore, the density of the universe was infinite. This discovery finally brought the question of the beginning of the universe to the realm of science.
>
> Hubble's observations suggested that there was a time, called the big bang, when the universe was infinitely small and infinitely dense. Under such conditions all laws of science, and therefore all ability to predict the future, would break down. If there were events earlier than this time, then they could not affect what happens at the present time. . . . One may say that time had a beginning at the big bang. (1988, 8–9)

Since Hubble's observations were made, evidence has continued to accumulate in support of the occurrence of this primal, cosmic event.

What sort of an event was the big bang? It was nothing quite like our experience with firecrackers and other terrestrial explosives (even atomic explosives) would lead us to expect. Suppose someone tosses a hand grenade into a crowded marketplace. The fiendish device goes off. There is a flash, there is a bang, and fragments fly hither and thither, doing untold damage to people and objects at various distances from the blast center. The big bang was not like this. Weinberg puts it this way: "In the beginning there was an explosion. Not an explosion like those familiar on Earth, starting from a definite center and spreading out to engulf more and more of the circumambient air, but an explosion that occurred simultaneously everywhere, filling all space from the beginning, with every particle of matter rushing apart from every other particle" (1988, 5).

How could this happen? Robert Wald, in his textbook on the general theory of relativity, observes that in the primal singularity state in which the universe began, the distance between all points of space was zero. It is not merely that the distance between objects was zero, but the very distance between the points of space was zero. What on earth does this mean?

To get at least some idea of what is going on here, we must examine some of the ways in which modern physics views such things as space and time. Consider two axes at right angles to each other, one labeled the *x-axis*, and the other labeled the *y-axis*. The axes can be imagined to extend from plus infinity to minus infinity. Taken together, they can be used to define a flat, two-dimensional surface known as a two-dimensional Euclidean space. Any point in this space can be denoted by a pair of numbers (x, y), one from the x-axis and one from the y-axis. If a third axis, perpendicular to the other two, is added and labeled the *z-axis*, the result is a three-dimensional space whose points are denoted (x, y, z).

The theory of relativity is formulated with respect to a four-dimensional geometric structure called *spacetime*. In this context, time is treated like a spatial dimension and is represented by an axis labeled as the *t-axis* perpendicular to the other three familiar axes. In real-world applications, spacetime is four-dimensional, and points are represented by quadruples of numbers (x, y, z, t). We humans did not evolve to do relativity theory (or advanced mathematics). Evolution has thus not equipped us to visualize four-dimensional objects in a four-dimensional spacetime (i.e., objects with length, breadth, width, and duration, or temporal extension). This does not impede a precise mathematical analysis of the properties of such objects. Moreover, it *is* possible to visualize events in a two-dimensional spacetime, where events are represented by pairs of numbers (x, t), one from the familiar spatial x-axis, and the other from the time axis (see Maudlin 2002, chs. 2 and 8; see also Ellis and Williams 1988; Shanks 1991, 1994).

In everyday life, we are used to thinking of space as something that is flat. What this means is that the shortest distance between two points, given by the Pythagorean theorem, is a straight line, that lines at right angles to a given line are parallel lines that never meet, that the internal angles of triangles add up to 180 degrees, and that $\pi = 3.14159\ldots$ represents the ratio of circumference to diameter for circles. But in relativity theory, geometries for spacetime are explored where these features do not hold. These geometries involve considerations of curved spacetimes.

For a simple example, consider what happens when you draw your x-axis and t-axis on the curved surface of a sphere with the origin $(0, 0)$ represented at a point on the equator (serving as the x-axis) that intersects with the t-axis represented by a line of longitude, which can then be thought of as the prime meridian. Notice that two lines of longitude, both at right angles to the equator, now meet at the poles, that the sides of triangles bulge out, and that the internal angles no longer add up to 180 degrees. (Try drawing a triangle on a sphere with one side represented by one quarter of the equatorial circumference and the other two sides represented by lines of longitude—perpendicular to the equator—that then extend to the North Pole in such a way as to be mutually perpendicular at the North Pole. The internal angles of this triangle consist of the sum of three right angles [i.e., 270 degrees].)

You will also notice that the ratio of circumference to diameter for circles drawn on the sphere is no longer 3.14159 . . . because the diameter of such a circle reflects the curvature of the surface on which it is drawn. Notice also that neither the x-axis nor the t-axis is infinite in extent. Follow either of them long enough, and you end up back where you started because of the curvature of the surface. (Straightest lines—shortest routes from A to B—on the surface of a sphere are represented by great circles.)

Finally, there is the matter of the curvature of surfaces. Since it is easier to think of circles drawn on paper than higher-dimensional surfaces, consider several such circles of differing sizes. The curvature of a circle is calculated as $1/r$, where r is the radius. Since smaller circles have smaller radii than larger circles, this formula tells us that smaller circles have greater curvature than larger circles. As the radius of a circle gets closer and closer to zero, the curvature gets greater and greater. At $r = 0$, it becomes infinite. (The same is true of spheres, where curvature is calculated as $1/r^2$.)

In relativity theory, the curvature of spacetime is something that reflects the distribution of matter and/or energy. Einstein's famous equation, $E = mc^2$, to be discussed shortly, tells us that energy and matter are different sides of the same coin, and because of this, physicists sometimes speak of mass-energy. Roughly speaking, the greater the concentration of mass-energy in a region of spacetime, the greater the curvature in that region. Locally, spacetime in the vicinity of the mass-energy of the Earth is curved, and the motion of the moon around the Earth reflects this curvature. On a larger scale, the motion of the earth around the sun reflects the greater curvature of spacetime brought about by the much larger localized concentration of mass-energy constitutive of the Sun.

In like manner, the distribution of mass-energy throughout the universe has global implications for the curvature of spacetime for the universe as a whole. Evidence from several quarters has indicated that our universe is expanding. It is now getting bigger, and thus it must have been smaller at earlier times. In turn, this means that mass-energy must thus have been more concentrated at earlier times. The curvature of spacetime must thus have been greater at earlier times.

Our universe has structure (planets orbit stars, stars belong to galaxies, and so on), and matter appears to be clumped on the small scales where you can, so to speak, see the individual trees in the wood. Nevertheless, if you take a bigger, global view of the universe and look at large scales, so you can see the wood as a whole, the distribution of mass-energy is very nearly uniform. Hogan has recently commented on the significance of these observations as follows:

> A long time ago, in the early Big Bang, the universe was uniform on much smaller scales, and even very small bits of it were flying apart; *very* early, even things a few inches apart were flying away from each other. Today, matter on small scales has congealed into stable systems that no longer expand, because over small regions, where the expansion is not too fast, forces have reversed the expansion. On these small scales, things are no longer uniform; matter is in stable "lumps" (galaxies and their contents), which are flying apart from each other but are themselves not expanding. (1998, 48–49)

We must now see how the current structure of our universe reflects its origins in more detail, for our universe is evidently evolving and changing with time. It is a dynamic structure.

In relativity theory, the effects of gravity are explained in terms of the curvature of spacetime and not, as they were for Newton, in terms of forces acting on objects in an otherwise flat, Euclidean spacetime. In regions where spacetime is curved by the presence of mass-energy, physicists sometimes speak of gravity wells. Black holes represent extreme cases, for in these regions mass-energy is so concentrated that the wells reflecting the curvature of spacetime have the property that not even light can escape from them. How could this happen? First consider the Earth. To escape from the Earth's gravity well, the curvature of spacetime in the vicinity of the planet is such that an escape velocity a little over 11 km per second must be achieved. This was achieved by the *Apollo* astronauts who went to the moon. In the case of black holes, the curvature is so extreme that the escape velocity exceeds the speed of light (300,000 km per second), so not even light can escape the gravity well, hence the name *black hole*. At the bottom of a black hole is a singularity. At a singularity, the concentration of mass-energy—hence the curvature of spacetime—is infinite. According to the big bang theory, our universe (mass-energy along with spacetime) emerged and continues to unfold from just such a singularity.

In the singularity state from which the big bang began, space itself would have had zero size, and so, as Wald has pointed out:

> The big bang does not represent an explosion of matter concentrated at a point of a preexisting, nonsingular spacetime, as it is sometimes depicted and as its name may suggest. Since spacetime structure itself is singular at the big bang, it does not make sense, either physically or mathematically, to ask about the state of the universe "before" the big bang: there is no natural way to extend the spacetime manifold and metric beyond the big bang singularity. Thus general relativity leads to the viewpoint that the universe began at the big bang. (1984, 99)

What unfolds and emerges from the primal singularity is spacetime itself—the very spacetime in which our Sun, our planet, and we ourselves would eventually come to be located. Until spacetime unfolds, there are literally no places (identified by coordinates x, y, z, t) for events and happenings to be located. There can be no history because there are no stretches of time for historical events to occur in.

A question that is often asked is this: What happened exactly at the time of the big bang (as opposed to a few microseconds afterwards)? Also, what happened before the big bang? It is well worth pausing to look briefly at the issues these questions raise. In the primal singularity state, when everything in the universe we see today was scrunched up into an infinitely small volume, cosmological theories, like the general theory of relativity, no longer work. Thus Ellis and Williams observe that at the point of the primal singularity "general relativity theory predicts a breakdown of space-time structure and known physics in the early universe in all realistic universe models. To investigate this further, one must move to a full quantum theory of gravity which has general relativity as a classical limit. The nature of such a theory is a problem which is still far from being fully resolved" (1988, 277). If our statements about the instant of the big bang are to have meaning, they must be formulated from a theoretical perspective that has at least roots in scientific evidence. But we currently do not have such a theory.

Physicists can extrapolate back to about 10^{-43} second after the big bang (the Planck time), which, though close, is not as far back as we would like to go. Craig Hogan explains the significance of this observation as follows: "The Planck length [10^{-33} cm] is as many times smaller than the width of a human hair as that width is smaller than the observable universe. The Planck time [10^{-43} sec] is the amount of time it takes light to travel that distance. This is the smallest interval of time and space; below it, the quantum curvature effects of gravity are so large that the notion of a simple, continuous spacetime becomes inconsistent" (1998, 7). Physicists have found it very hard to unite general relativity (the physics of the large and massive) with quantum theory (the physics of the very small).

This is a problem, precisely because, at the earliest times, everything that is now in the universe at large was scrunched up into something so small that quantum effects, already significant for objects of atomic size (10^{-8} cm) and smaller, would have been important effects. Yet at present we have no renormalizable quantum gravity theory. What this means is that we currently have no theory that yields physically meaningful numbers about the earliest times and smallest sizes of the universe (though the hunt is definitely on). Some physicists think superstring theories may help resolve these matters, but as of the time of writing, too many issues remain unanswered for there to be much confidence in the fruits of these lines of theoretical inquiry. Cosmology is very much a work in progress.

If to speak about the universe is to speak from the standpoint of an evidentially grounded theory, then it looks like the issue of the bang itself, and what happened before, is an issue where the philosopher Wittgenstein's advice is worth taking: "Whereof one cannot speak, thereof one must be silent." Hogan thus observes of the question as to what happened prior to the big bang:

> Compare this question with medieval speculations about what happens at the edge of the world. A believer in a flat Earth is faced with either an infinite world or one with an edge, whereas with a round Earth, the question of an edge ceases to have any relevance. Asking what came before the Big Bang might be like asking what is north of the North Pole—a place where "north" has no meaning. It is presumptuous to assume that just because they are suitable for talking

about the nature of time today, our ideas of time must also apply to the utmost extremities of spacetime. (1998, 56)

Once you are at the North Pole, all directions are south. Asking what happened ten minutes before the big bang is like asking for a point ten miles north of the North Pole. Stephen Hawking puts it this way: "The quantity that we measure as time had a beginning but that does not mean that spacetime has an edge, just as the surface of the Earth does not have an edge at the North Pole" (1989, 68–69). Asking what happened an hour before dinner might make sense; asking what happened an hour before the big bang does not.

Scientific theories rooted in evidence are our best guides. Without them, however imperfect they may be, there is nothing to say, nothing to be right or wrong about. In our discussion of the evolution of the human eye in chapter 2, we saw that in the land of the blind, a creature with a single photosensitive cell was king. So, too, in the land of ignorance. Here a partial theory is better than none at all. Even though we are far from having all the answers and there are significant gaps in our knowledge, 300 years of modern physics and cosmology has left us far from being blind about the universe we live in. Gaps in our knowledge are gradually being filled, even though gaps about the very beginning remain.

There is no doubt that our perspectives on singularities will develop in tandem with good cosmological science. And there, perhaps, is the point. Real physics does not rest content with invocations of supernatural beings with magical powers to solve puzzles. Genuine puzzles are recognized as such and become the subject for the next generation of theorists to contend with. These theorists, in turn, are constrained in their intellectual activities by the slow accumulation of data and theory. There is nothing like this developmental process in supernatural science, which is fixed and forever beyond evidence-driven revision and change.

Given the gaps in our current understanding of the universe, physics as we currently have it becomes possible only very shortly after the primal singularity begins to unfold. Within a fraction of a second of the big bang, the infant universe is very small, very curved in on itself, very dense, and very hot—as much as 10^{11} degrees Celsius (much hotter than the center of the Sun). The big bang initiates the expansion of the universe we live in, but the story does not end here. The initial expansion was amplified by something known as an *inflation mechanism* (see Guth and Steinhardt 1989 for details of the inflation concept). Hogan has recently observed:

> The expanding universe at the heart of this model is thought to be given its form by the action of an energy field, the *inflaton*. With the right properties, the inflaton's interactions can lead to repulsive gravity and create an instability that drives the original expansion of the Big Bang by making everything fly apart from everything else. We think this process of inflation is the way the universe got to be much bigger than an atom. Inflation is what made the Big Bang big. (2002, 453)

The inflation mechanism thus made an initially small universe huge. But a consequence is that the universe in the large reflects its origins in the small.

The inflaton-field is a quantum field. Like other quantum fields (e.g., the quantum electromagnetic field), it was not a smooth, uniform, continuous field, but one that contained random energy fluctuations that gave rise to inhomogeneities. These fluctuations were important because they have the consequence that the effects of inflation were not uniform. Some regions were affected more than others by the inflation mechanism. And so Hogan notes:

> The inflaton fluctuations are very important. For one thing, they are the reason the universe eventually broke up into galaxies, stars and planets. The inflaton fluctuations, frozen into the fabric of space, were converted into very slightly denser and sparser regions of matter. The denser regions eventually collapsed due to gravity. Without these perturbations, the universe would still be perfectly smooth today. Every galaxy we see (even whole clusters of galaxies) ultimately derives from about one elementary inflaton particle in the early universe. (2002, 424)

As the universe expands, it cools. After about 100,000 years, the universe would be cool enough for atoms of hydrogen and helium to form. Under the influence of mutual gravitational attraction, these primal gases would coalesce to form the first stars and galaxies (Weinberg 1988, 5–10).

In the hearts of these early stars, the conditions of extreme temperature and pressure were such that hydrogen atoms would fuse together to form helium atoms. As noted previously, one of the central equations of modern physics is Einstein's celebrated equation, $E = mc^2$, where E stands for energy, m for mass, and c is the speed of light (already a large number at 300,000 km/sec, its square is enormous). The equation tells us that mass and energy are different sides of the same coin. In fusion events, light atoms meld to form heavier atoms. In the process, small amounts of mass are converted into enormous amounts of energy in accord with the equation $E = mc^2$ (even relatively small amounts of mass multiplied by c^2 will yield relatively large quantities of energy). A similar fusion process takes place in so-called hydrogen bombs, where the extremes of temperature and pressure are generated by the detonation of an atom bomb as a trigger. The energy of the atom bomb comes from the fission (splitting) of heavy atoms—typically isotopes of uranium or plutonium—to form lighter atoms. In this process, too, small amounts of mass are converted into enormous amounts of energy.

Fusion events are nuclear events; that is, they concern the nuclei of atoms. As noted before, fusion is a physical process by means of which lighter nuclei can form heavier nuclei. From these early beginnings, events in the stars took the primal hydrogen and helium and began turning it into the heavier elements in the periodic table of elements. Importantly, carbon was produced in abundance—for we are carbon-based life forms. Ordinary stellar mechanisms will generate elements up to about iron in the periodic table. To get the heavier elements, such as lead, gold, and uranium, special stellar events known as *supernovas* must occur. These events are caused when large stars collapse in on themselves. A supernova is a massive explosion, and it can outshine an entire galaxy. The resulting energies are such that the truly heavy elements in the periodic table can form by the fusion of lighter nuclei. In a sense, then, all of us here on Earth are made of star stuff formed in the hearts of stars.

Cosmology and Thermodynamics

But what about the laws of thermodynamics discussed in chapter 3? Don't they require that we pay more attention to the issue of the origin of the universe, even if it does come from a primal singularity? Creation scientist Henry Morris once put it this way:

> Thus, the Second Law proves, *as certainly as science can prove anything whatever*, that the universe had a beginning. Similarly the First Law

> shows that the universe could not have begun itself. The total quantity of energy in the universe is a constant, but the quantity of *available* energy is decreasing. Therefore as we go *backward* in time, the available energy would have been progressively greater until, finally, we would reach the beginning point, where available energy equaled total energy. Time could go back no further than this. At this point both energy and time must have come into existence. Since energy could not create itself, the most scientific and logical conclusion to which we could possibly come is that: "In the beginning, God created the heaven and the earth." (Quoted in Strahler, 1987, 89)

Doesn't the very existence of the universe point to a supernatural creator? Isn't the very origin of the universe something that lacks a natural explanation?

Advocates of supernatural explanations for the origins of the universe are indeed apt to say that the universe was designed and created intelligently by an enormously powerful being. Since the time of St. Augustine, it has been popular in Christian circles to say that the creation was literally creation from nothing (creation ex nihilo). Notice first of all that if this is the case, the violation of the law of conservation of energy is only shunted back one level. And no explanation is given for how it is that the creator, by its supernatural constitution, manages to get something for nothing.

Such a supernatural solution looks like a cop-out. Invoking supernatural beings and supernatural causes (about which little is ever said and even less evidence is presented) amounts to little more than a shallow excuse for a violation of the laws of nature. When push comes to shove, advocates of this solution change the rules of the game. Magic is invoked, and laws are freely violated. With the explanation of creation ex nihilo, we might just as well have said Abracadabra!

We can do better than this. A number of physicists have addressed the question of the total energy of the universe from the standpoint of methodological naturalism (see Adair 1987, 364; Stenger 1998, 10; Rees 2001, 141–143). As noted previously, one of the central equations of modern physics is Einstein's celebrated equation, $E = mc^2$. The equation tells us that mass and energy are different sides of the same coin. In nuclear reactors and atomic weapons, for example, small amounts of mass are converted into large amounts of energy, whereas in particle accelerators, enormous amounts of energy can be used to form particles of very small mass. Martin Rees has observed that once gravitational energy is given due consideration in the settling of nature's energy accounts, it is possible for the net energy of the entire universe to be zero:

> Everything has an energy equal to mc^2, according to Einstein's famous equation. But everything also has negative energy because of gravity. We on Earth have an energy deficit compared to an astronaut in space. But the deficit due to all the masses in the universe added together could amount to *minus* mc^2. In other words the universe makes for itself a gravitational pit that exactly compensates for its rest-mass energy. So the energy cost of inflating our universe could actually be zero. (2001, 141–142)

Perhaps, then, the positive energy of matter is exactly counterbalanced by the negative potential energy of gravity. (Potential energy is energy an object has due to its position. A ball at the top of a hill has potential energy, and as it rolls down the hill, its potential energy is converted into kinetic energy of motion equal to $\frac{1}{2}mv^2$, where m is the mass and v is the velocity of the ball.)

If the net energy cost of inflating our universe is zero, nothing needs to be added from outside. The energy accounting books that are internal to the system balance. In these terms, it takes no energy at all to get a universe, and no thermodynamical work needs to be done from outside the system (see also Guth and Steinhardt 1989, 54). The books would be out of balance, and in need of a serious accounting adjustment from outside, if you had the positive energy of matter alone, created from nothing.

What then of the Second Law? Doesn't the claim that the universe is a closed system mean, in accord with the Second Law, that since the entropic disorder of the universe has been tending toward a maximum, there must have been an earlier time, the time of creation, when it was at an absolute minimum, when everything was perfect and ordered? As noted by Stenger, this argument does not take into account the observation that the universe is expanding:

> The second law argument holds only for a universe of constant volume. The maximum entropy of any object is that of a black hole of the same volume. In an *expanding* universe, the maximum allowable entropy of the universe is continually increasing, allowing more and more opportunities for order to form as time goes by. If we extrapolate the Big Bang back to the earliest meaningfully definable time, the so-called *Planck time* (10^{-43} second), we find that the universe started

> out in a condition of maximum entropy—total chaos. The universe had no order at the earliest definable instant. Rather than going from perfect order to disorder, the universe went from chaos to localized order. (1998, 10)

In chapter 3, we saw how open-dissipative systems could self-assemble and self-organize in accord with the Second Law to generate these local islands of order.

In other words, far from degenerating from an initially perfect, created condition (a sort of cosmic Garden of Eden), the universe started out messy and contains such order as we find thanks to self-organizing processes generated by mechanisms operating in accord with the laws of physics. As Hogan observes, "The evidence we have already suggests that the universe began with almost no information, and all the complex structure within it has developed since the beginning, on its own, without external influences" (2002, 427). Our universe is currently far, far away from a state of thermodynamical equilibrium, and our local corner provides us with many examples of the self-organization of ordered systems—systems whose existence is driven by the flow of usable energy from the Sun.

The Anthropic Coincidences

More sophisticated design arguments do not rely on misunderstandings about the Second Law. They rely instead on the so-called anthropic coincidences. The story behind these numbers takes us back to the 1920s and the work of Sir Arthur Eddington. (See Gale 1981, 1997; Gale and Shanks 1997 for information on the intellectual background against which these early debates took place.) Eddington, the leading mathematical physicist of his generation, was unhappy with the way that the laws of physics were rooted in observation and experiment. He felt that they should be deducible from pure mathematical theory. To this end, he took some of the basic numbers in physical theory, such as the charge on the electron and the mass of the proton, and started manipulating them to see if anything was awaiting discovery.

What Eddington found was that the ratios of these numbers came out to numbers of the order of magnitude of 10^{40} (regardless of whether the component numbers of these ratios were measured in volts, feet, pounds, and so on). Eddington never did find out what lay behind these enormous numbers. Though he tried to formulate a theory of the universe to explain these numerical coincidences, his colleagues remained nonplussed—harsh critics accusing him of dabbling in numerology and magic. This didn't make numbers of the order of magnitude 10^{40} go away, and attempts to explain what was going on persisted into the 1930, when one of the architects of the modern quantum theory, Paul Dirac, argued that the numbers might reflect the current age and structure of the universe and that the numbers might have been different at earlier times (and might be expected to be different at later times in the history of the universe).

In 1961, the American physicist R. H. Dicke observed that the ratio for a typical stellar lifetime to the time for light to traverse the radius of a proton has to be of the order of 10^{40} if stars are to generate carbon. He also showed that this ratio must be of the same order of magnitude as the ratio of electromagnetic force to gravitational force between two electrons if the universe is to contain elements heavier than lithium (Stenger 2000, 50). Dicke speculated that if all this was correct, then, given the biological importance of stuff like carbon, only certain numbers (the components underlying these mysterious ratios) might be suitable for the origin and evolution of life. The Greek word for "human" is *anthropos*. The numbers that made our kind of carbon-based life possible became known by the 1970s as the *anthropic* numbers, and since there are several such numbers, they are known collectively as the *anthropic coincidences*. With the coincidences in place, the stage was set not only for intelligent design creationism but also for some very interesting debates in theoretical physics.

The situation as it exists in contemporary cosmology with respect to these numbers has been explained by Stephen Hawking as follows:

> The laws of science, as we know them at present, contain many fundamental numbers, like the size of the electric charge of the electron and the ratio of the masses of the proton and the electron. We cannot, at the moment at least, predict the values of the numbers from theory—we have to find them by observation. *It may be that one day we shall discover a complete unified theory that predicts them all, but it is also possible that some or all of them vary from universe to universe or within a single universe.* The remarkable fact is that the values of these

numbers seem to have been finely adjusted to make possible the development of life. (1988, 125, my italics)

The cosmological variables whose values are not currently nailed down by physical theory, but whose particular values are essential for the existence of life on Earth, are the ones that have come to be known as the anthropic coincidences. As we shall see, merely naming them does not solve the interpretation of their significance, and in fact there is considerable debate over the significance of these numbers.

Before discussing this further, we must first be introduced to some of the numbers themselves whose existence generates the controversy. In the standard model of cosmology, there are some eighteen numbers that must be determined experimentally. Martin Rees has identified six particularly crucial numbers (1999, 2–3):

1. The number N, which measures the strengths of the electric forces holding atoms together divided by the force of gravity between them. N is a big number. If its value was much smaller, the universe would not exist long enough for life to evolve.
2. The number ε, which determines how strongly the components of atomic nuclei bind together. Its value influences the fusion processes in the hearts of stars and hence has an important effect on what chemical elements the universe contains. Its particular value helps explain why carbon and oxygen are abundant.
3. The number Ω, which determines the amount of matter in the universe (hence the importance of gravity). Too big a value, and the universe would have collapsed long ago; too small a value, and galaxies would never have formed.
4. The number λ, which governs a newly discovered force—a sort of cosmic antigravity whose effect influences the large-scale structure of the universe. A value much larger than the one we observe would have prevented galaxies (hence us) from forming. This number played an important role in the inflation mechanism discussed earlier.
5. The number Q, which determines the structure of the fabric of the universe. Inhomogeneities at the time of the big bang are subsequently amplified through physical mechanisms into cosmic structures such as stars, galaxies, and galactic superclusters. A smaller value of Q, and the universe would

be structureless; a larger value, and the universe would be dominated by black holes. Either way, we wouldn't be here.

6. The number *D*, which determines the number of spatial dimensions in our universe. The value of *D* is 3. The existence of our sort of life hinges on *D* being 3 and not 2 or 4, for example. (In relativity theory, time is treated as the fourth dimension, but it is not quite identical to a spatial dimension.

These symbols, N, ε, Ω, λ, Q and D, can be thought of as variables that could have taken values that were different from the ones they actually took in this universe. Had they done so, we wouldn't be here. Concerning the significance of these numbers, Rees comments:

> These six numbers constitute a "recipe" for a universe. Moreover the outcome is sensitive to their values: if any one of them were to be "untuned," there would be no stars and no life. Is this tuning just a brute fact, a coincidence? Or is it the providence of a benign Creator. . . .
>
> It is astonishing that an expanding universe, whose starting point is so "simple" that it can be specified by just a few numbers, can evolve (if these numbers are suitably "tuned") into our intricately structured cosmos. (1999, 4)

I deliberately left part of this quotation out, because the traditional debate has focused on the choice between blind chance and intelligent design. As Rees himself observes (in the bit I left out), other options are possible. Chance or design is a false dilemma!

So far, then, we have physical laws that we humans have intelligently designed to make sense of the universe we live in. The laws contain symbols whose values are not fixed by the laws themselves but by observations we have made of the world around us. The values these symbols take are crucial for the existence of life as we know it. Currently, we do not know if we can come up with a better theoretical account of our universe in which the values taken by these symbols would be explained by the laws embodied in the theory. Though at present we have no theory that explains the actual values of the cosmological symbols that make us possible, we do have theories about the significance of these numbers. These theories are captured by a variety of anthropic principles, some more deeply rooted in science than others.

Anthropic Principles

In a now classic discussion of the anthropic coincidences, Barrow and Tipler (1986) offer a variety of statements that might serve as candidates for an anthropic principle in terms of which the so-called anthropic coincidences might be understood. The anthropic coincidences do not come labeled with explanations of their importance and meaning. This must emerge from theorizing. Unfortunately, speculative theorizing, not necessarily of a scientific nature, is what lies behind some of the anthropic principles (see Gale 1997).

In the discussion due to Barrow and Tipler (1986, 17–23), it emerges that there are three species of anthropic principle, under whose headings variations on common themes might be gathered: the weak anthropic principle (WAP), the strong anthropic principle (SAP), and the final anthropic principle (FAP). These terms can be explained as follows:

The Weak Anthropic Principle (WAP)

The observed values of all physical and cosmological quantities are not equally probable, but they take on values restricted by the requirements that there exist sites where carbon-based life forms can evolve and by the requirement that the universe be old enough for it to have already done so.

This principle captures the fact that we live in a universe that contains life (you and me). Any theory that explains the properties of the universe (including, perhaps, the anthropic coincidences) should be constrained by the requirement that it must be consistent with this fact about life. Those theories that are not consistent with this fact of life can be disregarded at the outset. Many scientists think WAP is a reasonable constraint on theorizing.

The Strong Anthropic Principle (SAP)

The universe must have those properties that allow life to develop within it at some stage of its history.

This statement is much more controversial. For scientists, the universe is a contingent thing. This means that though it came out this way as a matter of fact, things could have been different. The word *must* in the statement of the principle means that things could not have turned out differently. In some sense, life (you and me) was necessitated at the dawn of time. Scientists are very wary of claims like this. For this reason, SAP is quite controversial.

The Final Anthropic Principle (FAP)

Intelligent information processing must come into existence in the universe, and, once it comes into existence, it will never die out.

This has all the problems of SAP, with some additional ones on top. It is one thing to say life must come into existence and quite another claim to say that intelligence must come into existence (and once in existence, never die out).

Perhaps because they were both fully aware of concerns about FAP and SAP, Barrow and Tipler caution:

> We should warn the reader once again that both the FAP and the SAP are quite speculative; unquestionably, neither should be regarded as well-established principles of physics. In contrast, the WAP is just a restatement, albeit a subtle restatement, of one of the most important and well-established principles of science: that it is essential to take into account the limitations of one's measuring apparatus when interpreting one's observations. (1986, 23)

Stenger (2000) puts WAP even more plainly: "If the universe were not the way it is, we would not be here talking about it."

If WAP is uncontroversial, FAP is most assuredly not. (Indeed, some commentators have called it the **C**ompletely **R**idiculous **A**nthropic **P**rinciple.) Though Barrow and Tipler said little about it, Tipler has recently published what he takes to be a scientific proof of the resurrection of the dead (1995). The aim of this book, which can only be described as a grotesque science fantasy presented as fact, is to argue that we have a reasonable hope, rooted in science, of life after the death of the body. The basic idea is that our personalities or souls are merely programs running on the hardware of our brains, and though our bodily hardware may perish: "The dead will be resurrected when the computer capacity of the universe is so large that the amount of capacity required to store all possible human simulations is an insignificant fraction of the entire capacity" (1995, 225). The reader interested by these ideas is warned that they are buried in more than 500 pages of extremely dense prose, the willing penetration of which requires a certain degree of psychic and spiritual desperation. I will not have anything further to say about FAP, which in any case does not figure significantly in the debates about the anthropic coincidences. The same cannot be said for SAP.

Barrow and Tipler offer three possible interpretations of the SAP, and these have come to set the contours of the contemporary debates (1986, 22):

(A) There exists one possible universe "designed" with the goal of generating and sustaining "observers."

(B) Observers are necessary to bring the universe into being.

(C) An ensemble of other different universes is necessary for the existence of our universe.

I will consider the implications of these interpretations in turn.

First of all, option (B) does not figure significantly in contemporary debates about intelligent design cosmology, where the arguments focus on options (A) and (C). It is tempting to see option (B) as something whose interest is restricted to New Age mystics, solipsists, and a small number of modern adherents of Bishop Berkeley's philosophy that to be is to be perceived or to be a perceiver. Finding solipsism unintelligible and New Age mysticism (along with Berkeley's speculations) to be silly, I am personally inclined to regard (B) as a version of (A).

But things are not quite as simple as I have characterized them, and physicists such as John Wheeler have given option (B) some serious consideration (see Gale 1997), primarily because some interpretations of quantum mechanics give the conscious observer a special role in the characterization of measurement interactions. That conscious observers play such roles is highly controversial (see Shanks 1993, 1995). For good or ill, my focus is thus on options (A) and (C).

Universe by Design

That there is but one universe, finely tuned by an intelligence at its origin for the emergence of creatures like us, has enormous appeal. It is here that the faithful have sought the stamp of science for their religious beliefs. Versions of this argument have recently been developed by Tipler (1995), Miller (1999), Bradley (2001), Corey (2001), and Ross (2001).

In typical presentations of this cosmological design argument, we are given a choice of two options. The first option is that our presence in the universe is the result of blind chance. Pure luck gave the cosmological variables their values, and here we are. Life is a happy coincidence. The other option is that the constants were intelligently tuned at the beginning, it is no accident we are here, and it is the result of deliberate design. In discussing this dilemma, we should bear in mind that we may have left out important options. There is no proof that there are only two options to choose from.

Moreover, even if we conclude that the constants were tuned by design, the numbers themselves tell us nothing about who or what did it. To get from the conclusion of mere design to the further conclusion that it was design by the God of religion X will require more than simple observations of cosmological fine-tuning. Such strong conclusions will involve (among other things) debates about the interpretations of sacred texts and philosophical and theological speculation of a kind that has no place in science.

Kenneth Miller, who has defended biological evolution from creationist attacks and has objected vociferously to claims made by Michael Behe, has made the leap from fine-tuning to the God of Christianity. Miller's version of this providential tale begins this way:

> Having decided to base life on the substance of matter and its fine tuned properties, a Creator who had already figured out how to fashion beauty and order . . . could easily have saved his greatest miracle for last. Having chosen to base the lives of His creatures on the properties of matter, why not draw the origins of His creatures from exactly the same source? God's wish for consistency in His relations with the natural world would have made this a perfect choice. As His great creation burst forth from the singularity of its origin, His laws would have set within it the seeds of galaxies, stars, and planets, the potential for life, the inevitability of change, and the confidence of emerging intelligence. (1999, 252)

Miller, needless to say, does not enlighten his reader as to how he knows all this, though he does note (253) that the "ultimate purpose of the work of God may never be understood by the mind of man." I can only conclude that Miller, who is a well-known professional biologist, wants creationists out of his own backyard and is happy to see them dumped on hapless physicists and cosmologists!

Nevertheless, what are we to make of this choice between intelligent design and chance? The natural tendency is to say that surely the anthropic coincidences couldn't have happened just by chance, from which, if you buy into the dilemma, you are led by the nose to the conclusion of design. It will not go amiss to examine issues surrounding both design and chance in a bit more detail.

I will begin with the issue of design. There is a tendency in these discussions to simply grant supernatural design or creation as a possibility. But it is far from obvious that passive acceptance of the hypothesis that a supernatural agent could design a universe (perhaps by initiating a big bang while simultaneously tweaking the cosmological parameters to get the universe in which we are to be found) even makes sense, let alone is true or false. While we know a lot about physical objects and how physical objects interact with each other, we know nothing about supernatural objects and how (if at all) they can interact with physical objects.

The situation here raises issues similar in some respects to old debates about the relationship of mind to body. Some people (they are known as Cartesian dualists) think that the mind is made up of a different substance from the body. In particular, the body is viewed as a physical substance, while the mind is viewed as a nonphysical, spiritual substance existing independently of physical nature—a substance, moreover, that can survive the complete destruction of the physical body. (For this reason alone, the theory has attracted the interests of those given to certain types of theological speculation.)

But no good account has ever been offered as to how a nonphysical substance can interact with a physical substance (so that nonphysical thoughts about food may lead the physical body to the refrigerator). Since our minds clearly influence our bodies, an alternative possibility (of interest to science) is that the mind, far from being a different substance from the body, is in point of fact a neurophysiological phenomenon—a phenomenon dependent on brain processes. On this view, minds are parts of physical nature, and to study the mind, one needs to focus on physical objects such as brains. This approach has yielded much scientific fruit. Neurophysiologists have managed to identify regions of the physical brain responsible for many cognitive functions and mental processes. That the mind should be understood in terms of the physical brain is a conclusion bolstered by studies of mental impairment arising from changes in the brain due to accidents, disease (e.g., Alzheimer's), or drugs. Mental illness, once viewed as a sign of supernatural demonic possession, is now viewed as something with treatable physical causes.

Yet, just as we have no good account of how a nonphysical mind could interact with a physical body, we have no account whatsoever of how nonphysical, supernatural beings could interact with a physical world. The cosmological intelligent design claim would be interesting if we had a demonstration that there are supernatural beings (or at least some good reason to think this was the case) and that some of these beings actually had the wherewithal to design a universe like ours. We could then ask whether such a being, or a collective of such beings, did design the universe, as a matter of fact. But we have nothing beyond the assurances of the faithful that talk of supernatural beings is both meaningful and factual.

Intelligent design theorist William Dembski believes that the design and creation of the universe was done with a spoken word. He observes, "Intelligent design, on the other hand readily embraces the sacramental nature of physical reality. Indeed, intelligent design is just the Logos theology of John's Gospel restated in the idiom of information theory" (2001b, 192). How does he know this Logos theology ("In the beginning was the Word. . . .") is correct? Where is the evidence? That it is in the Bible is no guarantee that it even makes coherent sense, let alone is a candidate for truth or falsity.

Before we can ascertain whether something is true or false, we must minimally ensure that the claim is meaningful. That natural human designers can make pocket watches and mousetraps tells us absolutely nothing about the intelligibility of talking of supernatural beings, still less about whether there are supernatural beings as a matter of fact, and whether it makes sense to suppose they *can* design natural universes—all matters that need to be settled before asking whether in fact one such being actually did design our universe.

Suppose you read in a science fiction novel, "Captain Shanks, daring commander of the space-cruiser *Darwin*, was hauling his mass out of a gravity well, escaping from hordes of slimy, slippery Idiots from the dreaded planet Id, when he hit the accelerator on the snagglefarg drive, thus warping his ship into Jabberwocky space and traveling with infinite speed to the Sombrero galaxy." It might make for a good tale if you don't think it through. But there is absolutely no reason to believe that what has been written here is even meaningful, much less that it describes a real possibility that may be a candidate for truth or falsity. In a similar way, if the choice is between chance and design, where talk of supernatural design is incoherent babble—even comforting, incoherent babble—chance wins every time! The flat refusal of design theorists to even try to spell out the details of their supernatural design scenarios only adds to the suspicion that there is much, much less here than meets the eye.

What now of chance? Suppose I take a deck of cards that has been shuffled very well. I tell you I am going to lay out the cards one after the other, dealing from the top of the deck. I ask you to guess the sequence in which I will lay them out. You write down your guess, and then I deal out the cards. No doubt we soon find that your guess and the actual sequence differ. There are 52 cards in the deck, and the chance that you will guess the exact sequence is 1 in 52! ($52! = 52 \times 51 \times 50 \times, \ldots, \times 1$). This is approximately equal to 1.2×10^{-68}—a very small chance indeed. Though you almost certainly couldn't guess the sequence in advance, once I have laid all the cards out, the probability that the sequence is that particular sequence is 1 (or 100%). The deck was shuffled fairly; any one sequence was as likely as any other. And it is worth adding that merely knowing that the probability of getting a given sequence of cards in a given trial, for example, is 1 in 52! tells you nothing about where, in a run of many trials, you will get that particular sequence. (If you roll a die, the probability of getting a 3 is 1 chance in 6 or 1/6. This doesn't convey information about which particular roll of the die, in a run of several rolls of the die, will yield a 3.)

The first lesson is that we should not confuse probabilities of events that have already occurred (we are here in the universe today) with probabilities of events before they occur (who could have guessed at the moment the big bang took place, before the cosmic deck is dealt, that we would be here today?). Unlikely events *do* happen, just by chance. This is one of the reasons why the best laid, intelligently designed plans of mice and men so often go astray, especially when they turn probabilities reflective of their ignorance of how various states of affairs may be realized into probabilities for the actual occurrence of those states of affairs.

You and I are here today. Never mind the universe and the anthropic coincidences, for the moment. Just consider how improbable it is that you are situated where you currently are. For example, growing up in Manchester, England, I could never have imagined I was going to finish up as a professor in upper eastern Tennessee writing a book about intelligent design theory. Think of all the things that must have gone right (and perhaps wrong) for this to have happened (all the cars that missed me while I crossed the road, all the planes that didn't crash, all the illnesses I didn't get), all the things that didn't happen to my parents (especially in World War II) and to my grandparents, and so on. Yet here I am, and so are you!

With respect to the universe, we are told that the probability for anything other than intelligent design is vanishingly small. If the analysis is rooted in data, a different conclusion emerges. As noted by Stenger (1998, 2000), the number of observed universes, N_o, is 1. The number of observed universes with life, N_l, is 1. The probability that any universe has life, based on observed data, is given by $N_l/N_o = 1$ (i.e., 100%). The statistical error is, needless to say, large.

There are other probabilities that need to be carefully differentiated. Stenger (1998) observes that we should not confuse the probability that one universe selected at random from a set of all possible universes would be our particular universe with the very different probability that a universe selected from this set of all possible universes could sustain some form of life (maybe very different from our own carbon-based forms). While the value of the first probability is no doubt vanishingly small, for all we know, the value of the second might be close to 1 (i.e., 100%).

First, other forms of life may be possible with different values of the cosmological variables, even if our form of life is not. (Stenger [2000] explores this possibility with the aid of computer simulations.) No demonstration to the contrary has been offered anywhere. Second, in other universes, where not just the cosmological variables but the very laws of physics and chemistry might be different, we have no reason to believe that our expectations about what is possible with respect to life forms in this universe carries over. We do not know, and, importantly, neither do design theorists. The argument from improbability is not even well defined unless we know how many of the alternative universes can support some form of life. If half the possible universes can support some form of life, then though selecting our particular universe at random may be unlikely, selecting a universe with life in it would be an event with a probability of .5 (50%). But we just don't know.

Rees (1997, 242) uses the following analogy, which he got from a philosopher called John Leslie. Suppose you are standing before a firing squad made up of fifty soldiers. They shoot; they all miss. Because they miss, you are here to tell the tale and, importantly, to wonder why. Perhaps it is just a matter of chance? This is not very satisfying as an explanation. Perhaps it calls out, then, for an explanation in terms of intelligent design. Perhaps it was an elaborate conspiracy to scare the dickens out of you.

But let's play a game known as misanthropic roulette. Like the Russian version, it is played with a revolver; unlike the Russian version, all but one of the six chambers are loaded. When you play the game, you spin the cylinder on the revolver, put the barrel to your head, and pull the trigger. What is your chance of surviving if you play the game? You have a chance of 1/6 (i.e., 16.7%) of surviving and 5/6 (i.e., 83.3%) of dying. What is your chance of playing twice and surviving to tell the tale?

Your chance of surviving the second round is independent of your chance of surviving the first round. If the second round is played, your chance of survival in that round of the game is 1/6. In this case, to calculate your chance of surviving both rounds of the game, we must multiply the probabilities of surviving each round separately. The probability of surviving both rounds is thus $1/6 \times 1/6 = .028$ (i.e., a little less than 3%). Suppose you decide to play the game five times. Your chance of surviving all five rounds is $1/6 \times 1/6 \times 1/6 \times 1/6 \times 1/6 = .00013$ (i.e., about 1 chance in 10,000). While this is a small number indeed, and you might decide not to play the game in the light of this information, this sequence of outcomes is as likely as any other sequence of outcomes, all of which involve at least one bullet being fired. The only difference is that a sequence of five clicks on an empty chamber is a sequence in which you survive to tell the tale. That you might, anthropically, care about this sequence of five clicks does not change its likelihood of occurrence, nor, if you play the game five times and survive, does it mean that the hand of providence had intervened (notwithstanding all the rabbits' feet, four-leaf clovers, and other assorted charms you may have placed into your pockets before playing the game).

One hard-headed response to the fine-tuning arguments—and I think it is the right response, for what that is worth—is that it was just a matter of luck, and maybe not so hard-headed, if the only alternative should turn out to be incoherent. We know unlikely events do happen. We have no reliable evidence for the existence of a supernatural cosmic universe-tuner, except as an explanation for what might be attributed to luck. This latter, however, seems to be little more than a cosmological version of the gambler's fallacy, manifesting itself here in the urge to offer causal explanations for the lucky streak of coincidences we had with the values of the cosmological variables.

Yet the explanation of coincidences is a big part of the issue. Thus intelligent design theorist Bruce Gordon has recently written, "The most intuitive explanation for this incredible string of coincidences, is, of course, design." He adds, "So far, the scientific community has rejected such a response as being outside the pale of science because it is interpreted as violating the canons of methodological naturalism" (2001, 203). It is a fact that many scientists (though not all, by any means) reject design because it is perceived as involving a craven appeal to the supernatural without adequate supporting evidence (other than to explain coincidences).

Gordon goes on to suggest a naturalistic design hypothesis that he thinks might enable us to overcome these worries:

> Perhaps our universe is embedded in another, much larger, physical universe and exists as the result of an experiment conducted by highly intelligent embodied beings who live in this larger universe. Of course, this only pushes the design problem back one step: where did *their* universe come from, and what are the conditions that made *it* possible? To avoid the specter of design altogether, a thoroughly naturalistic account of the origin of all possible physical universes, and of our own in particular, would have to be devised. (2001, 203)

This is highly reminiscent of Behe's willingness to consider alien designers (as opposed to supernatural designers) in the context of biochemistry, and so is Gordon's problem as to who designed the designers.

Luckily, design, be it natural or supernatural, is not needed—that is, mandated by the available evidence. Some cosmologists have indeed considered the possibility that our universe might be part of a larger structure—but they have done so without appeals to intelligent designers. These theoretical views need to be discussed not to show that design hypotheses are false but to show that there are other naturalistic explanations of the same facts that lead design theorists to postulate a supernatural designer. The choice may not be simply a matter of one universe by chance *or* one universe by design.

Multiple Universes and Meta-Cosmology

Earlier in this book, we saw that considerations rooted in natural theology have led some theorists to suppose that the natural universe is the fruit of supernatural intelligent design, so considerations rooted in natural meta-cosmology have led some scientists to see the natural universe as an undesigned part of a larger undesigned structure whose properties render our anthropic coincidences less mysterious.

To see what is going on here, consider what happens when, instead of having just one person play five rounds of misanthropic roulette, we have many millions or billions of people trying to play five rounds. On the first round, we will lose 5/6 of the players, but 1/6 will go on to play the second round. Of these survivors, 1/6 will survive to play the third round, and so on. The more players who start out, the more players there will be who, by chance alone, survive five rounds of the game. This leads us to a consideration of Barrow and Tipler's option (C): the multiple-universes option (see Gale 1990 for a taxonomy of theories about multiple universes).

Some modern cosmologists are currently investigating the concept of the multiverse. This is the idea that our universe may just be one of many universes. Rees explains the idea as follows:

> There may be many "universes" of which ours is just one. In the others some laws and physical constants would be different. But our universe would not be just a random one. It would belong to an unusual subset that offered a habitat conducive to the emergence of complexity and consciousness. The analogy of the watchmaker could be off the mark. Instead, the cosmos may have something in common with an off-the-rack clothes shop: if the shop has a large stock, we are not surprised to find one suit that fits. Likewise, if our universe is selected from a multiverse, its seemingly designed or fine-tuned features would not be surprising. (2001, 164–165)

On this view, our universe, which we have traditionally thought of as everything, is really a mere part of something much, perhaps infinitely, bigger.

The big bang, far from being a special and unique creative event in which everything begins, might only be an event of relative insignificance in a much larger structure. Our laws of physics, far from being universal, might be mere local bylaws. The multiverse hypothesis does to the anthropic universe what Copernicus's heliocentric hypothesis did to the cosmological vision of the Earth as a fixed center of the universe. Since the time of Copernicus, it has emerged that we live on one planet orbiting an insignificant star in an insignificant galaxy. The hypothesis before us is that our universe is one (potentially insignificant) universe among many. Perhaps we can out-Copernicus Copernicus!

The hypothesis of a multiverse, which would offer a naturalistic explanation of the anthropic coincidences, is certainly worthy of consideration. For one thing, it suggests that the old dichotomy of chance or design is not quite right. An obvious objection springs from something known as *Occam's razor*, named after the medieval philosopher, William of Occam. Occam's razor, as it appears in scientific debates, is the principle that we should not multiply entities unnecessarily. In these terms, the introduction of a multiverse to explain the anthropic coincidences smacks of a liberal application of what might be termed *Occam's hair-restorer*.

It might be argued, however, that the design hypothesis (assuming it even makes sense) is at least as hairy as the multiverse hypothesis, since both hypotheses explain the anthropic coincidences through the invocation of additional entities. Explanation by the hypothesis of one universe and blind chance or luck then emerges as the smart and clean-shaven hypothesis. This is certainly my view. However, the multiverse hypothesis does have one advantage over the design hypothesis in that it at least accords with our intuitions about the sorts of things—physical objects—that exist. Postulating a supernatural designer (assuming it even makes sense to do so) requires the introduction of a new type of object and causality into science: supernatural causes and supernatural objects. These latter are matters we know absolutely nothing about.

We may as well be clear about this: There is no independent evidential warrant for the postulation of nonphysical, supernatural objects whose only role is to serve as components of an (as yet unformulated) alternative explanation of phenomena for which a (religiously unpalatable) naturalistic explanation is possible. At least this latter kind of explanation manages to appeal to the kinds of objects—physical objects—for which there is independent evidential warrant. In these terms, perhaps we would be better off with the devil we all know—physical objects and physical causes—than the one we don't know.

And in science we have good evidence of the benefits that can flow from multiplication of entities belonging to categories for which there is independent evidential warrant—for example, physical entities. From the time of the ancient Greeks until the late eighteenth century, water was thought of as a homogeneous simple substance—an element. In the late eighteenth century, Henry Cavendish showed it was complex. It was not an element but composed of other physical substances, hydrogen and oxygen. Our thinking about steam under pressure enabled thermodynamicists in the first half of the nineteenth century to think about the steam engines that were powering the industrial revolution. To really get a good grip on thermodynamics, as we saw in chapter 3, required multiplying entities again, to see steam as something made up of molecules in motion. In eighteen grams of water, we now think there are about 6.02×10^{23} molecules of water. (Molecules are small. The same number of grains of sand would cover the entire state of Texas to a depth of more than thirty feet.)

Rees himself draws a parallel between this debate about Occam's razor and an earlier debate in the history of astronomy concerning the shapes of planetary orbits. It was traditional in astronomy to view orbits of celestial objects as being circular. The ancient Greek philosopher Plato had asserted that the circle was the perfect figure, and an obsession with circles can be found in the work of Aristotle and later thinkers sharing this intellectual inheritance. Even Copernicus, who placed the Sun at the center of the universe, retained the idea of circular orbits—as did Galileo. However, Kepler, using data gathered by Tycho Brahe, realized that the orbits were in fact elliptical.

No less a figure than Galileo himself was shocked by this. Yet Kepler was right. Thus Rees observes:

> Our Earth traces out one ellipse among an infinity of possibilities, its orbit being constrained only by the requirement that it allows an environment conducive for evolution (not getting too close to the Sun, not too far away). Likewise our universe may be just one of an ensemble of all possible universes, constrained only by the requirement that it allows our emergence. So I'm inclined to go easy with Occam's razor: a bias in favor of "simple" cosmologies may be as shortsighted as was Galileo's infatuation with circles. (1999, 156)

Rees continues:

> If there was indeed an ensemble of universes, described by different "cosmic numbers," then we would find ourselves in one of the small and atypical subsets where the six numbers permitted complex evolution. The seemingly designed features of our universe shouldn't surprise us, any more than we are surprised at our particular location within our universe. We find ourselves on a planet with an atmosphere, orbiting at a particular distance from its parent star, even though this is a very "special" atypical place. A randomly chosen location in space would be far from any star—indeed, it would be in the intergalactic void millions of light-years from the nearest galaxy. (1999, 156–157)

As things stand, our universe has a beginning in the big bang. If there is a multiverse, the beginning of our universe, which we call its creation, might really be merely a process of change occurring in part of this preexisting, larger structure (where our ideas of space and time would not apply). This structure might have always been there, undesigned and with no beginning, undergoing complex processes of change.

But in the end, there is always the question of evidence. Intelligent design theorists tell us nothing about the designer, save that they think it ought to be the God of Christianity. The methods and materials employed by the designer and any account of supernatural objects themselves (how they differ from physical objects, how they bring about effects in the physical world) are apparently beyond the scope of human knowledge. I argued in the last chapter, in my discussion of the intelligent design questions, that without some honest account of these matters, the invocation of a supernatural cause for any particular thing we currently don't understand, let alone the universe itself, is tantamount to mumbling the magic word *Abracadabra*. Intelligent design theorists do not even have a program of research in which answers to the intelligent design questions might meaningfully be sought.

And what of the hypothetical multiverse? Alas, neither I nor anyone else has a Julesvernoscope that would give us a glimpse of the multiverse by night. Some theorists such as Rees (2001) think that the multiverse hypothesis might be testable. But such tests as he envisions are highly indirect. The multiverse hypothesis might better be viewed as a piece of naturalistic metaphysics, by contrast with the intelligent design hypothesis, which is clearly nothing more than supernatural metaphysics.

Though highly speculative, the multiverse hypothesis is an alternative to the hypothesis of one designed universe and the hypothesis of one universe simply due to chance alone (my preferred view). It is at least conceivable that there are features in our own universe that might carry information about the multiverse itself (or hypotheses that are formulated concerning its characteristics). In this sense, the idea of a multiverse may not be entirely beyond the scope of human ken. It frankly remains to be seen whether the idea of a multiverse will do useful work in science.

At rock bottom, however, it is a fact that the cosmological numbers are not fixed by current physical theory and have to be empirically determined. Perhaps other values of these variables are possible. This is not currently known. One certainly cannot infer from the fact that the values of these variables are not fixed by current physical theory that they could have taken any value whatsoever, other than those they did take as a matter of fact in our universe. Maybe they could; maybe they couldn't. We don't know. Nothing follows about the nature of reality (and the ways it could have been different) from our ignorance of it.

It is possible that one day we will discover a theory that predicts the values taken by these cosmological variables in our universe. Of this possibility, Rees comments:

> If the underlying laws determine all the key numbers uniquely, so that no other universe is mathematically consistent with those laws, then we would have to accept that the "tuning" was a brute fact, or providence. On the other hand, the ultimate theory might permit a multiverse whose evolution is punctuated by repeated Big Bangs; the underlying physical laws, applying throughout the multiverse, may then permit diversity in the individual universes. (1999, 157)

Either way, we have no such theory now. There is currently no rationale, rooted in our current best physical theories, for the postulation of supernatural intelligent design of the universe. It remains at best an evidentially ungrounded possibility—a possibility from which rational beings may reasonably withhold their assent. At worst, it may be incoherent nonsense and hence not even a candidate to be judged right or wrong. We do not know, and advocates of supernatural intelligent design have evidently decided to say nothing helpful.

Conclusion

In the course of this chapter, we have seen that even amid the uncertainties of modern cosmology it is not a simple matter of choosing between the hypothesis of one universe resulting from creation and design and one simply due to chance. The multiverse hypothesis is as deserving of consideration as are these two traditional hypotheses. Moreover, while it is at least arguable that the multiverse hypothesis and the creation-design hypothesis are both metaphysical in the pejorative sense of that word, it is far from obvious that the creation-design hypothesis is sufficiently well formulated to be deemed coherent or possible, let alone likely or indeed preferable to an explanation of the same facts as either being due to chance alone or to events in a multiverse. Once again, we have not been presented with the extraordinary evidence that would warrant a leap to the conclusion of intelligent design. At various points in this book we have seen how the misleading appearance of intelligent design can be generated by the combined effects of chance and dumb mechanisms operating in accord with the dumb laws of nature. There is currently no evidential reason to believe that the appearance of cosmological intelligent design is anything more than an illusion explicable in terms of purely natural effects.

Conclusion

Intelligent Designs on Society

So far in this book, we have traced the origins of the contemporary intelligent design movement, and we have seen how Darwin's theory of evolution and its subsequent development in the long course of the twentieth century has completely reshaped our understanding of life on Earth, including our own estimate of our place in nature. We have seen that the emergence of order and complexity of the kind exhibited by organisms is quite consistent with the laws of thermodynamics and, indeed, that the processes of self-organization described in the context of nonequilibrium thermodynamics explain complexity and order in both living and nonliving systems. It was against this background that we turned to assess the arguments of contemporary intelligent design theorists.

We saw that coming out of antiquity were two versions of the argument from design. One was biological, and the other was cosmological. Contemporary advocates of intelligent design have attempted to revive both of these arguments and, in the course of their analyses, to call into question the naturalistic assumptions upon which they claim modern science is rooted. But we have seen that it is necessary to differentiate metaphysical naturalism from methodological naturalism. Adherents of the latter, but not the former, can entertain hypotheses about the supernatural and about supernatural intelligent design in particular. The methodological naturalist will, however, based on long experience in which only natural objects have shown themselves for scientific study, demand high-quality evidence before giving serious consideration to hypotheses about the supernatural. The metaphysical naturalist, by contrast, will merely ignore and dismiss claims about supernatural objects and their causes and effects.

But we also saw that there is a position that I termed *methodological supernaturalism*. The methodological supernaturalist is inclined by faith to accept supernatural claims, but this theorist also recognizes the need to produce high-quality evidence to get claims about the supernatural a fair hearing among those who lack faith in the existence of such objects but whose view of the world is sensitive and responsive to the production of evidence. The common interest in evidence means that there can be a certain overlap of interest between methodological naturalists and methodological supernaturalists, even though their initial orientations are very different.

Contemporary advocates of intelligent design, however, have failed to take the evidential high road, preferring instead to refute parodies of natural science while refusing to face their own evidential obligations. To a person, they have refused to offer credible evidence that there *are* any supernatural objects. They have also failed to explain how supernatural objects can interact with physical objects. They have not explained how we might even begin to gather evidence to examine claims about the details of design. Was there one designer or a committee of designers? Did they exploit quantum fluctuations, or was it ripples in ectoplasm? We don't know, and they are unwilling to tell. It is sometimes said that nature abhors a vacuum. If so, it must abhor contemporary musing on intelligent design, for such musings, though undeniably voluminous, are evidentially empty.

We have examined the claim that organisms exhibit a kind of complexity known as *irreducible complexity*, which supposedly could not possibly be explained in terms of natural evolutionary processes. We saw this claim was simply untrue. We also saw that the redundancies in biochemical and molecular systems that science has actually uncovered provide a conceptual framework in which the phenomenon of irreducible complexity can be explained without recourse to designers of unknown origins, using unknown methods and materials for unknown purposes.

We also examined the claim that contemporary cosmology provides evidence in support of supernatural design and creation. While questioning the intelligibility of this claim, we saw that there were alternatives that explained the same data, and that did so without recourse to unknown denizens of a supernatural domain unknown to science. The conclusion, once again, was that we had been given no serious evidential case upon which to base further investigations into supernatural intelligent design. The late Carl Sagan advised that extraordinary claims require extraordinary evidence for their validation. Intelligent design advocates have not merely failed to offer extraordinary evidence but indeed have failed to offer even humdrum evidence to support their case. In fact, intelligent design theory, for all its blather about being the science of the twenty-first century, is little more than old medieval theological wine in new biochemical and cosmological bottles.

There is one last issue to discuss, and this concerns matters of value rather than matters of fact. A hypothesis to be considered at this juncture, especially given the contrast in content between claims made at the thin end of the wedge and claims made at the fat end, is that intelligent design theory is not a genuine attempt to advance scientific knowledge after all. It is in fact an ideological vehicle carefully crafted to advance a preexisting, conservative Christian social agenda. These matters are important, not least because there are many Christians who are not conservatives and because there are many conservatives, as well as Christians, who do not share the narrow social agenda and hostility to science reflective of the extremist fragment of Christianity under analysis here.

Designs on the Body

My curiosity about these matters was piqued by the metaphysical speculations of political scientist John West Jr., a senior fellow of the Discovery Institute (the home base for intelligent design theory). In a discussion of the implications of intelligent design theory for questions about the science of the human mind, West observed:

> If intelligence itself is an irreducible property, then it is improper to try to reduce mind to matter. Mind can only be explained in terms of itself—like matter is explained in terms of itself. In short, intelligent

> design opens the door to a theory of a nonmaterial soul that can be defended within the bounds of science. At the very least, if intelligence is understood as an irreducible property of human beings, the grounds on which science can undercut free will and personal responsibility will be significantly diminished. (2001, 66)

But this appears to be profoundly muddled. Issues of free will and responsibility are logically independent of possession (or not) of a nonmaterial soul.

West seems to confuse materialism (according to which only physical objects exist) with some species of determinism (according to which free will is an illusion). Moreover, there is no reason to believe that a nonmaterial soul could not shackle your will as much as physical brain processes are alleged to do (ignoring all the well-known attendant problems of explaining an interaction between the physical body and the nonphysical soul). Free will is not guaranteed simply through an appeal to the nonmateriality of the soul. However, we soon learn that West has little interest in trading metaphysical niceties.

It turns out that the real issues motivating West's ill-advised foray into the realm invisible, with a view to illuminating the freedom of the will, turn out to be welfare reform and programs that benefit the poor. He informs us: "As long as human beings are regarded as the helpless victims of material forces, it is understandable why welfare policies would focus narrowly on changing material inputs rather than also looking at issues of character and accountability" (2001, 66). However, there is absolutely no reason why a materialist, as opposed to a certain kind of determinist, might not think character and accountability are important, too, and important, moreover, not only for the poor (for whom this talk of character and accountability is often a prelude, in right-wing Christian circles, to assigning them blame for their wretched condition) but also in the boardrooms of corporations, where greed and decisions freely made concerning the maximization of profits often have the consequence that many ordinary, decent people require welfare assistance (through no fault or defect in their nonmaterial souls, let alone their characters or their failure to be accountable for their actions).

West goes on to tell us: "Human beings know intuitively that they are more than a bundle of cells and synapses" (2001, 67). Of course, human beings are more than a pile of cells and synapses. Humans, like all organisms, are complex, organized systems that exist and function because of myriads of interactions, internally between cells and synapses and externally with the environments in which they are embedded. Arguably, then, organisms have properties as wholes that cannot be reduced to the properties of the parts. But it doesn't follow from this that they possess supernatural parts, such as nonmaterial souls. Moreover, when it comes to treating neurological disorders such as depression, Parkinson's disease, stroke, and Alzheimer's disease, not to mention other diseases such as diabetes, heart disease, and a veritable host of neuromuscular disorders, the scientific study of those same cells and synapses is of crucial importance. Metaphysical musings about nonphysical souls will only impede our understanding of these matters.

Even if it should turn out that mind itself is not simply reducible to matter, it does not follow that mind is independent of matter and thus, in West's words, "can only be explained in terms of itself." The enormity of the ignorance of neurobiology in West's claim is so great as to almost defy comprehension. Antidepressants (and other drugs used in the treatment of mental illness) have given relief from psychological distress to many thousands of people. One type of matter (the drug) affects another type of matter (the evolved brain). Relief from psychological pain and distress (phenomena of the mind) is the result.

In a similar vein, it is merely disingenuous for West's colleague Phillip Johnson to claim: "The parts of biological science that are achieving real progress would be only modestly affected by the recognition that intelligent causes necessarily played a crucial role in biological creation" (2000a, 168). We have seen in this book that there is no evidential warrant for the recognition of intelligent causes. The parts of biology where progress is being made would have to abandon the very high standards of inquiry that have made them successful in order to embrace intelligent causes on the basis of the arguments, taken singly or as a group, that are constitutive of intelligent design theory.

Such a gruesome embrace of supernatural causation would inevitably come at the expense of evolutionary biology. Yet evolutionary biology is the veritable glue that holds all the disparate branches of biological inquiry together and gives common focus to their collective endeavors. Take away evolution, and the biological sciences would degenerate into an incoherent collection of rudderless ships. Given the role played by evolutionary considerations in human biomedicine, these are not risks to be taken lightly (see Ewald 1994; Nesse and Williams 1995). Johnson's additional claim that the parts of biological science where progress is being made "are only nominally committed to materialist assumptions" (2000a, 168) is sufficiently laughable as to merit no further comment here. What does he think cells and molecules are made of?

Those who seek to advance real biomedical science need to study human cloning, stem cells, and human embryonic tissue. The current religiously motivated, political restrictions on these forms of scientific research in the United States are a national disgrace. People will get sick, they will suffer, and they will die for want of politically restricted research. However, the issues are not merely restricted to the future of scientifically meaningful research programs. Some more immediate issues of great social interest are also at stake. West tells us:

> A third area where intelligent design has cultural implications is the sanctity of human life. As noted previously, the argument over eugenic abortions and assisted suicide is premised largely on the notion that human beings are the sum of their material parts. If human beings are more than this, then the cogency of this argument disintegrates. Once the idea of a nonmaterial soul gains new currency, the ethical context in which issues such as abortion and euthanasia are debated will considerably expand. (2001, 67)

This is simply false. Arguments in favor of abortion or assisted suicide have nothing to do with humans simply being the sum of their material parts. For that matter, mere possession of a nonmaterial soul doesn't seem to be particularly relevant either. There is no guarantee that such an object, were it to be discovered, would have the properties attributed to it by conservative Christians (many of whom seem strangely dedicated to the vigorous application of the death penalty).

In fact, the important ethical issues about abortion have little to do with magical beliefs about nonmaterial souls and a lot to do with the ways in which, as reasonable, compassionate human beings, living in a pluralistic society where people are still free to have wide ranges of religious and nonreligious moral beliefs, we are to care for victims of incest and rape, to care for sick mothers, and to respect the informed choices of women facing unwanted pregnancy. These issues are often further complicated by lack of ready access to contraception and to information and education about family planning. None of these issues can be reduced down to bumper sticker slogans such as RU-486 or

RU-4Jesus? or Choose Choice. They are really tough ethical issues, a sensible treatment of which will only be impeded through the interjection of religious beliefs that interested parties may reasonably deny.

Similarly, the ethical issues about assisted suicide have a lot to do with the ways in which we should respect the autonomous choices of the terminally ill and of those seeking to escape the unrelenting pain that is associated with some forms of chronic illness. The denial of access to such relief, on the basis of unfounded magical beliefs about souls that others may reasonably not share, strikes many of us who have cared for sick loved ones, dying by degrees in incredible pain, as not merely lacking in compassion but evil beyond belief. All these ethical issues are complex and will admit of no easy resolution, one way or the other. But it is very clear that an important part of the aim of intelligent design theory, as an alternative to science as we know it, is to provide justifications, under cover of smoke and mirrors, for antecedently held positions on abortion and assisted suicide that are dear to religious extremists.

Designs on Morality

One does not have to scratch far below the surface to get much closer to the real motivations of the intelligent design movement, which in reality have little to do with science but a lot to do with politics and power—in particular, the imposition of discriminatory, conservative Christian values on our educational, legal, social, and political institutions. The real issues are not about science but about who shall count, whose views shall be heard, and who will be silenced. This is the dark end of the wedge, and it is more than a little disturbing.

Thus, according to Nancy Pearcey, a senior fellow at the Discovery Institute, her colleague Phillip Johnson has recorded that he has encountered fears that if naturalistic evolution is discredited, women will be sent back to the kitchen, gays back to the closet, and abortionists to jail. Pearcey then goes on to comment:

> Though the fears Johnson encounters are certainly exaggerated, *the basic intuition is right*, for the question of our origin determines our destiny. It tells us who we are, why we are here, and how we should order our lives together in society. *Our view of origins shapes our understanding of ethics, law, education—and yes, even sexuality*. If our life

> on earth is a product of blind, purposeless natural causes, then our own lives are cosmic accidents. There's no source of transcendent moral guidelines, no unique dignity for human life. On the other hand, if life is the product of foresight and design, then you and I were meant to be here. In God's revelation we have a solid basis for morality, purpose and dignity. (2001, 45, my italics)

The real issues in this debate about intelligent design thus have a lot to do with a rejection of the democratic values of the Enlightenment and a desire to reimpose, by using appeals to the supernatural, a narrow and restrictive Christian morality onto people—very often traditionally oppressed minorities—whose very freedom and self-expression depend on their having thrown off those same restrictive and discriminatory values.

When proponents of intelligent design direct their criticisms at modernists, it is very clear that it is the heirs to our common Enlightenment heritage that they are talking about. Pearcey is nevertheless right about one thing. How one sees one's origins and place in nature can have a profound influence on one's views about ethical, social, and political matters. An extreme example is afforded by the case of the Church of the Creator, a white supremacist church that promotes a version of Christianity that embraces the idea of the battle of good against evil as involving a race war of whites against nonwhites. There are valuable lessons here for those interested in how myths about origins can be used to legitimate preexisting antisocial behaviors and prejudicial beliefs and attitudes.

Moral Theory and Moral Practice

It is often in the context of these divisive, vexatious debates that religious conservatives claim to take the moral high road. Yet the road traveled by religious zealots is rarely the one they claim to be on. If we look at what radical Hindus, Islamic fundamentalists, extremist Jews, and conservative Christians actually do, rather than what they say we should all do, a disturbing divergence between moral theory and practice emerges. Extremist religious belief has played a significant causal role in much of the bloodshed and slaughter in human history. Its role in perpetuating cruelty, misery, and terrorism is not hard to find in the modern world.

In our own culture, Christ's worthy message of the Sermon on the Mount is rarely, if ever, put into practice by those who claim to speak for Christ. More than a century ago, John Stuart Mill noted in *On Liberty* (1859), in a discussion of Christian moral principles:

> Yet it is scarcely too much to say that not one Christian in a thousand guides or tests his individual conduct by reference to those laws. The standard to which he does refer it, is the custom of his nation, of his class, or his religious profession. He has thus, on the one hand, a collection of ethical maxims, which he believes to have been vouchsafed for him by infallible wisdom as rules for his government; and on the other a set of everyday judgements and practices.... To the first of these standards he gives his homage; to the other his real allegiance. (1972, 101)

The moral problems involved in this divergence between moral theory and actual practice run deep because the problems are inextricably intertwined with corrupt and twisted practices on the part of religious authorities themselves—the alleged worldly authorities who claim to speak for God and the absolute moral standards that are to guide our behavior.

This is not the place to parade pedophile priests (see Boston Globe Investigative Staff 2002) or twisted televangelists. The problems here are not new. Geoffrey Chaucer, writing in the fourteenth century, long before the birth of Charles Darwin and the rise of modern science so resented by intelligent design theorists, saw the problem clearly:

> And this figure he added eek thereto,
> That if gold ruste, what shall iren do?
> For if a preest be foul, on whom we truste,
> No wonder is a lewed man to ruste;
> And shame it is, if a preest take keep,
> A shiten shepherde and a clene sheep,
> Wel oghte a preest ensample for to yive,
> By his clennesse, how that his sheep should live.
> (Winny 1965, 66–67)

A neglected moral issue in this debate about values is evidently manifested in the divergence between the theory and practice of a distressingly large number of those very authorities who speak the moral message of Christianity.

It is in fact far from clear that complex ethical matters should be left in the hands of various religious authorities, simply because they are religious authorities. There is at most a casual connection between religious belief and ethical behavior. Moreover, real ethical dilemmas are usually sufficiently complex to resist treatment in terms of platitudes and slogans about what God does and does not will, and about what little we can know of his plans for us. It is no use saying that we cannot be good without the absolute moral standard provided by God, for this absolute moral standard has been freely used to justify of all manner of atrocities. An honest catalog of actual human acts sanctioned by appeals to the absolute moral authority of God would no doubt reveal the deity to be an abhorrent monster.

And (borrowing a turn of phrase from Wittgenstein), lest our thought goes sick on an unbalanced diet of examples, recall that among those who call themselves Christians and act on behalf of their faith are not just the likes of Mother Theresa and the ordinary decent folk who, without fanfare or media attention, perform simple acts of kindness and do the best they can to be good people. We will also have to consider the approximately 70,000 members of the Church of the Creator, mentioned previously, which boasts branches in 48 states and 28 countries. This is a church whose followers "have shot, knifed or beaten blacks, Jews and Asian Americans" (Nicholas Kristof, *New York Times*, August 30, 2002) and which promotes racial holy war on behalf of whites. The actual moral message of Christianity is a mixed one, not only in theory but also in practice.

If Christians are to be placed on the scales of good and evil, they must all be so placed, not just those who have performed good works. When properly loaded, the scales will balance, on one side, a few conspicuously saintly folk and a larger number of decent, ordinary people struggling to get by as best they can and, on the other side, a motley crew of sadists, racists, and bigots ranging from godly members of the Ku Klux Klan and the Aryan Nations to the divinely inspired perpetrators of some truly heinous acts. (Recall that Adolf Hitler ended chapter 2 of *Mein Kampf* by observing, "Hence today I believe I am acting in accordance with the will of the Almighty Creator by defending myself against the Jew. I am fighting for the work of the Lord.") I can think of no better practical reason for separation of church and state than the existence of this mixed moral bag, consisting of the good, the bad, and the ugly, and all claiming to speak for God.

Of course, it may well be replied at this point that at least Christians know what they ought to do, even if they often fail in practice. But many contradictory voices speak sincerely and authoritatively about what Christians do and do not know about these matters, especially when it comes to particular moral cases where there may be genuine disagreements over which general principles must be applied to fit novel situations and where there are deep divisions concerning how those principles are to be interpreted and rendered meaningful (for example, in contexts such as the death penalty, the decision to go to war, care for the environment, care for animals, abortion in the context of rape or incest, or even simply helping the poor and needy). It is far from clear exactly what it is that all Christians know.

God and Value

Nevertheless, leading proponents of intelligent design are concerned that having a view of the world that does not deliberately make a place for God must involve having a valueless, amoral view of the world. Thus, cosmological creationist Kenneth Miller observes: "This—and not any departure from the discipline of science—is what distinguishes a believer from a non-believer. To a believer, the world makes sense, human actions have a certain value, and there is a moral order to the universe" (1999, 258). Yet this does not seem quite right. Is it really true that there can be no value without God? For example, is it true that there is no sense in talking of things getting better or worse without God? Most Christians seem to be capable of talking of the best make of SUV, the best neighborhood to live in, and the worst stocks to buy without any need to invoke the almighty as a standard relative to which these matters of value are to be judged. If these judgments of value—of what is a good choice to make and what is a bad choice to make—can proceed without God, why not other judgments of value?

This matter becomes all the more acute when we consider the radically differing estimates of what God actually stands for in the culturally diverse community of Christians. This is a community that does not speak with one voice. (Even in the more restricted community of creationists, there are enormous disagreements over the very date of creation itself—consider that young Earth creationists and old Earth creationists offer wildly differing estimates of the age of the Earth and hence the need to change science to suit religious purposes.)

Here I would like to examine this diversity in the community of Christians in more detail. It is a fact that Christians do not speak with one voice. The community of Christians is a complex one that reflects the diversity of the communities and cultures in which Christians find themselves. In the part of Tennessee where I live, for example, you can find Baptists who maintain that Catholics are not even Christians. Christians disagree with each other over what God is, how the Bible is to be understood, and what, exactly, God commands us to do. (Even the Ten Commandments have to be interpreted to be applied in particular cases.) These disputes are reflected in profound disagreements over what is moral and what is not. Once we get beyond sectarian squabbles over who has the true or the real access to God's plan for humans, it becomes a good question as to how we are to adjudicate moral disagreements. It is no use appealing to what God wants in order to settle these issues, since exactly what God wants is the very issue at hand.

For Christians reading this book, this issue of diversity is really important. The choice is not one between Darwinism and a narrow, exclusionary conservative conception of Christianity. There may be no absolute choice to make because the churches that contribute to Christianity's diversity have a wide range of opinions concerning Darwinism, with many notable churches being fully reconciled to the discoveries of modern science. Moreover, the freedom to have an opinion of your own is a characteristic of Protestantism. Arguably, there exists a plurality of opinion within the evangelical branches of Christianity. Conformity to the narrow strictures and conservative, exclusionary theology of the advocates of intelligent design is simply not an essential part of being a Christian, as many of Christianity's practitioners understand their faith.

Science and Morality

Michael Behe himself has argued that post-Darwinian science, bereft of a place for intelligent design, bequeathed a view of the universe as "a cold, bland place, indifferent to human life" (1999, 11). Citing research on the anthropic principle and fine-tuning arguments in cosmology, Behe adds that the universe has become "much cozier" (1999, 11). Now enhancing a sense of coziness is no part of the business of science, however attractive it might be to those who yearn for a fool's paradise.

And of course, even if the universe was designed—let us suppose, contrary to fact, that the cosmological design arguments actually worked—it would not follow that the universe was a cozy, warm, fuzzy place, a sort of intelligently designed Disneyland writ large. It could be designed and indifferent to human life. There is, moreover, no reason to believe that the alleged cosmological designer was a moral being. For all we know (and the problem of evil that worried Darwin—the existence of extensive suffering and misery in the world—is surely relevant here), it is a cruel joke or a malicious experiment. These sorts of design arguments, even if they were adequate in showing the bare fact of design, would leave us light years away from Michael Behe's optimistic hopes for a cozy cosmos.

How, then, can one be good without God? What is needed is a rational discussion that reflects the best estimates of the relevant facts as we have them, not one rooted in myths and fairy stories about the supernatural. The first step is to have a clear view of the sort of creatures we are and where we stand in nature, especially with a view to our relationships with other creatures. To this end, I think that Charles Darwin had the seeds of some good ideas in chapter 3 of his book *The Descent of Man*. These seeds may now be sown.

A Darwinian approach to morality looks not for transcendent supernatural values and an alleged moral order in the universe—an absolute moral order concerning which the faithful routinely disagree with each other, often to the point of slitting each other's throats. Rather, the Darwinian approach to morality begins by asking about the nature of the human condition. Given facts about the human predicament, it then goes on to ask what function morality serves. What is its adaptive value? What do our moral institutions do for us? As Darwin was aware, these questions require an understanding of human nature from the standpoint of both biological and cultural factors. So where has evolution left us?

To deal with this question, we will have to remain tightly focused on human beings and the sorts of creatures we are, as parts of nature. Here I run the risk of being accused of trying to infer what we ought to do from statements about what is the case as a matter of fact. (This is also an issue for those who would say, "God forbids X as a matter of fact; therefore, we ought not do X.") Those partial to slogans will say,

"You cannot infer an *ought* from an *is*." In fact, the relationship between facts and values is a complex matter that is by no means as easily settled as proponents of the *is-ought* slogan would have you believe. To short-circuit this controversial and vexatious issue, I note here that the factual information I will discuss will not logically determine what we ought to do as evolved beings capable of moral behavior, but it will nevertheless be morally relevant to discussions of what we ought to do. Facts (e.g., humans are soft, squishy creatures who are easily injured) can be relevant to discussions of value (e.g., one ought not drink and drive) without logically determining what we ought to do.

As Darwin was aware, nature affords examples of cooperation (bees in their hives, termites in their mounds, elephants in their herds, not to mention symbiotic relationships between members of different species) as well as competition. Evolution is pure demographics. It is about the production of offspring who can themselves reproduce. Ecological circumstances exist in which cutthroat competition is a way to succeed in this endeavor, but circumstances also exist in which cooperation with one's fellow creatures and even altruism can be as effective a strategy to the same end. (See Dugatkin 1997 for a good review of the literature on the evolution of cooperation and altruism.)

In this light, having an accurate view of oneself as part of nature will be a component of the wisdom resulting from the injunction to know thyself. Since no person is an island, this will necessarily mean knowing oneself in relation to other people and the rest of nature relevant to one's life and well-being. This will be the context in which we can ask what morality is for.

In this spirit of asking what morality is for, I note that Darwin himself had some useful comments on these thorny topics:

> As man advances in civilization, and small tribes are united into larger communities, the simplest reason would tell each individual that he ought to extend his social instincts and sympathies to all members of the same nation, though personally unknown to him. This point being once reached, there is only an artificial barrier to prevent his sympathies extending to men of all nations and races. If, indeed, such men are separated from him by great differences in appearance or habits, experience, unfortunately shews us how long it is before we look at them as our fellow creatures. (1871, 96)

Later in the same chapter, building on these insights, Darwin charts a possible course of human moral development as follows:

> But as man gradually advanced in intellectual power and was enabled to trace the more remote consequences of his actions; as he acquired sufficient knowledge to reject baneful customs and superstitions; as he regarded more and more not only the welfare but also the happiness of his fellow men; as from habit, following on beneficial experience, instruction, and example, his sympathies became more tender and widely diffused, so as to extend to men of all races, to the imbecile, the maimed, and other useless members of society, and finally to the lower animals—so would the standard of his morality rise higher and higher. (1871, 99)

Moral progress and development is thus reflected in the extent to which people in various communities and societies have been able to extend their limited sympathies and affections beyond the immediate domain of their primary social interactions with kin, extended families, and local communities.

Morality makes it possible for us to benefit from sociality and cooperation when affections and sympathies, especially those that might bind us to close kin, leave off. Diffusion of our sympathies and affections may expand the target population for social and cooperative interactions, but such diffused sympathies may lack the ability to bind us together. Here is the place for morality. It is a cultural device to amplify and extend the benefits of sociality and cooperation.

At the very least, our natural capacity for morality enables us to counter the natural limitations of our sympathies for others—for example, distressed strangers in our midst or others far removed from us geographically. It does this, not necessarily by making us have feelings about them (though that may happen), but by making us simply not indifferent to their plight and condition, even if we ourselves are not emotionally moved by it. We want good things for those we care about—those who are in the reach of our sympathies—and as moral philosopher Geoffrey Warnock has observed:

> If one can be (as most can be) moved *sometimes* by the predicament of another, then it is possible for one to want human predicaments to be in general ameliorated, and thus to feel as practically efficacious that range of reasons which has that ultimate rationale. . . . We may thus say that just as the *need* for morality, its having a point, derives from very

> general facts about human beings and their predicament, so also it is a fact about human beings that they are capable of morality; the disease and its remedy, so to speak, have a common source. (1971, 165–166)

We do not need supernatural intelligent design to have a moral nature. We need look no further than to the sorts of creatures we are and to the sort of world in which we are embedded. It is as much a part of our nature not to be indifferent to the plight of others generally as it is to be caring and compassionate toward those more clearly in the domain of our direct acquaintance.

Most of us are raised by social animals—our kin—who by their interactions with us participate in the developmental realization of our social inclinations, sentiments, and affections. They also reinforce the tendencies nature has given us. We are cognitively sophisticated social animals. The ethologist Konrad Lorentz once remarked: "Our categories and forms of perception, fixed prior to experience, are adapted to the external world for exactly the same reason as the hoof of the horse is already adapted to the ground of the steppe before the horse is born and the fin of the fish is adapted to the water before the fish hatches" (quoted in Cziko 1995, 73). Not just our categories and forms of perception are adapted, though, but our behavioral inclinations as well. We are ready to be social animals when we are born. The result of inheritance and socialization is the existence of beings who are capable of forming cooperative alliances with other people who are not even closely related. There is a large literature on the genetic mechanisms underlying the evolution of social behavior, especially cooperative behaviors in humans and nonhuman animal species (for we are not the only social, cooperative organisms to be found in nature). Ridley (1998) explores the consequences of some of these ideas for morality.

My preferred approach to these matters draws its inspiration from the writings of Peter Geach, a very conservative Catholic philosopher and thus an extremely unlikely source for someone like me to draw inspiration from, especially since Geach is opposed to Darwinian ideas. Geach observed some years ago that it makes sense when dealing with moral issues to ask, "What are humans for?" (by analogy with the question "What are eyes for?"). This is a great question to ask from a naturalistic Darwinian perspective, as well as from Geach's preferred theological perspective. It is a good Darwinian question precisely because evolutionary processes have shaped our nature behaviorally and psychologically and not simply morphologically. To answer the question "What are humans for?" you must know the nature of the beast.

Most of us would agree that it is in our nature to have social and cooperative inclinations, and these inclinations and their consequences when realized in action are important in accounts of the sort of creatures we are. Indeed, we might agree on this while differing on a wide range of other matters. In this regard, Geach commented:

> Consider the fact that people of different religions or of no religion at all can agree to build and run a hospital, and agree broadly on what shall be done in the hospital. There will of course be marginal policy disagreements, e.g., about abortion operations or the limits of experimentation on human beings. But there can be agreement on fighting disease, because disease impedes men's efforts towards most goals. (1979, 13)

A hospital is an example of a large-scale cooperative enterprise. It is not hard to come up with other examples where people of diverse opinions nevertheless manage to cooperate for mutual benefit.

It is in the context of such cooperative enterprises that we can see why human beings need the traditional moral virtues of prudence, justice, temperance, and courage. As Geach observed:

> These virtues are needed for any large-scale worthy enterprise, just as health and sanity are needed. We need prudence or practical wisdom for any large-scale planning. We need justice to secure cooperation and mutual trust among men, without which our lives would be nasty, brutish and short. We need temperance in order not to be deflected from our long-term and large-scale goals by seeking short-term satisfactions. And we need courage in order to persevere in face of setbacks, weariness, difficulties and dangers. (1979, 16)

We not only need these virtues to achieve the benefits of sociality and cooperation but also are, as a matter of fact, disposed by varying degrees to manifest them in our behavior. It is in our evolved nature. One does not have to look far to see that cultures and societies differ with respect to the ways in which they encourage or impede the developmental realization of our social and cooperative inclinations.

Indeed, each of the cardinal virtues can be thought of as convenient folk descriptions for that which can otherwise be thought of functionally in terms of clusters of related behavioral and psychological tendencies and dispositions. The virtues have been discussed since ancient times not because they reflect absolute moral standards derived from some supernatural realm but because they capture important features of our common evolved natures. It is upon these biological foundations that cultures are built and evolve.

The virtues, being rooted in our biology, are manifested like most other traits, with some considerable degree of variation. This variation ranges from the selfless saint at one extreme to the sociopathic moral parasite at the other. But the moral parasite can prosper only because most of us are not like him. That most of us are not like the moral parasite is why we can hope to prosper in turn by being social, responsible, and cooperative. We are capable of bettering the predicament in which we find ourselves and others. This is one of the great benefits of a capacity for culture and cultural evolution.

Thus, to be what we are for as the result of evolutionary processes that have bequeathed to us capacities for sociality and cooperation, we need virtues. But Geach has rightly pointed out:

> Men need virtues as bees need stings. An individual bee may perish by stinging, all the same bees need stings; an individual man may perish by being brave or just, all the same men need courage and justice. It is . . . sophistical to write as if the alternatives were: moral virtue for its own sake, and selfishness. Men are so made that they do care what happens to others; quite apart from respect for Duty, that is the way men's inclinations go. (1979, 17)

Cognitive sophistication has its place here. Imagination is very important. We are able to imagine ourselves in the shoes of others. We have a capacity for empathy. We can, of course, do to others what they have just done to us, but we are able to imagine others doing to us what we have done to them and, importantly, what we have imagined we did to them. Imagination can thus forestall deadly encounters and behaviors injurious to others. Put this way, the basis for morality is to be found in human beings as parts of nature. In particular, the basis of morality is to be sought in the characteristics of the relationships humans have with each other. These relationships are not fixed and unchanging. They can be modulated for good or ill. Moreover, if the function of morality is indeed the amelioration of limited human sympathies, we can make other relevant observations.

Being an unbeliever—seeing oneself as a natural part of a natural universe bereft of supernatural beings—does not preclude one's having a meaningful life reflective of decent and humane values. Neither does a denial that one lives in a universe intelligently designed with us in mind. To explain how life can be meaningful and purposive, we do not need to look beyond ourselves. Meaning and purpose come in no small measure from the relationships we forge with others. The others we relate to may be family members, friends, folk in the neighborhood, or people in the community and workplace. Meaning is also derivative in part from our relationship to nature itself. It is no accident that so many of us have and cherish pets or yearn for beautiful countryside and are horrified at how some of our fellow humans are destroying it—fouling our common nest—for the sake of short-term profits. People can be good, loving, and caring without God, and many of us are, as a matter of fact. We are social animals, and it is in our nature. Living in a world that is indifferent to us does not mean that we have to be indifferent to each other or to the rest of the world in which we find ourselves.

Evolution may help us understand why we have a capacity for morality, but it does not determine unique answers to specific questions about what we ought to do. This is the place for the hard moral discussions that we cognitively sophisticated creatures are capable of. The contours of these debates will certainly be shaped by many factors, including the cultural context in which they occur. A conscious awareness of the cultural factors that shape the contours of our moral debates, especially our social, political, and religious beliefs, can at least sometimes enable us to move beyond the divisive and destructive clutches of these same cultural factors.

Social animals enjoy many benefits from their sociality. Animal societies are self-organizing, complex systems composed of many interacting parts. This is as true of human societies as it is of those of ants and bees. Humans differ from ants and bees with respect to cognitive sophistication. The complex dynamical interactions that weld us together in our home communities, as well as the broader societies and cultures to which we belong, have not been left as our ancestors bequeathed them. Our intelligence—and in particular its action with respect to our capacity to be moved by considerations relating to the welfare of others—confers on us the ability to modulate the effects of self-organization. We can modulate, that is, the dynamical processes governing the interactions of individuals that generate and sustain our social lives—for both good and ill. When those interactions are modulated so as to promote the survival and well-being—in short, the flourishing—of individuals constitutive of ever more inclusive interacting groups, we have modulated them for good.

Individual moral development from a Darwinian perspective is the process by which an individual comes to extend his or her sympathies beyond the narrow confines of the self to the family, the community, and ultimately to society and the world at large. The moral development of a community or a society can be measured in part by the space its institutions create for the moral development of the individuals in it. In these terms, the extension of limited sympathies to include marginalized genders, marginalized races, and those marginalized by lifestyle choices others disapprove of (including not only groups marginalized by sexual orientation but also marginalized religious groups) is a sign of moral development.

This Darwinian position is not the same as moral anarchism, in which, from the standpoint of behavior, anything goes, regardless of how harmful or destructive it is. Nor is it the same as moral relativism, according to which all behaviors are equally good or equally justifiable. For example, the willful limitation of human sympathy to encompass a narrow range of interests or a narrow range of people who are otherwise indistinguishable, morally speaking, from those excluded will indeed be a mark of moral retardation.

What I have said here of broader social and political institutions applies with vigor to religious institutions themselves. Some religious institutions—for example, those engaged in charitable work for the poor—have promoted moral development by countering the natural limitations of human sympathies, but others have not. We see this in the religious strife in Northern Ireland, in Palestine, and in India and Pakistan. The extension of our limited human sympathies, which means, among other things, learning to tolerate and to live peaceably with those we disagree with, is an integral part of the democratic inheritance of the Enlightenment. Our long and bitter experience in the twentieth century shows that communities and nations forcibly welded together through the exclusive enforcement of a narrow range of interests, be they religious or secular, are unhealthy communities and nations. Their peoples may live together, but they surely do not flourish and bloom.

As for Darwin himself, he was a humane individual who even considered the extension of our limited sympathies beyond the boundaries of our own species:

> Sympathy beyond the confines of man, that is, humanity to the lower animals, seems to be one of the latest. It is unknown by savages, except towards their pets. How little the old Romans knew of it is shewn by their abhorrent gladiatorial exhibitions. . . . This virtue, one of the noblest with which man is endowed, seems to arise incidentally from our sympathies becoming more tender and more widely diffused, until they extend to all sentient beings. As soon as this virtue is honored and practiced by some few men, it spreads by instruction and example to the young, and eventually becomes incorporated into public opinion. (1871, 96–97)

For us, the result has been seen in the establishment of animal shelters, humane societies, and laws preventing cruelty to animals. We have come a long way since the time of Darwin. Rachels (1991) explores some of the important moral issues here. Perhaps the next step will be a realization of the importance of preserving and caring for the broader biological environment in which all sentient beings evolved and which will be needed if we are to flourish down the generations and not simply get rich today. This would be one good way in which we could all really care for those as yet unborn.

Science is clearly relevant to discussions of value. And so, as we confront difficult ethical questions in the context of biotechnology (ranging across contraception, abortion, genetic screening, cloning, and stem cell research, for example) and as we confront issues about environmental degradation, having a good understanding of evidentially grounded natural science is likely to be of great importance in arriving at rational decisions in both public policy contexts and our personal lives about what does and does not promote the amelioration of our limited human sympathies.

Religion and Reason

Phillip Johnson has been clearer than most about the depth of the issues at stake here in this debate about the moral implications of the intelligent design movement. He has recently observed: "The

Enlightenment rationalists thought that it was safe to reject God because it seemed to them that human desires are basically rational, and so the human problem is simply to choose the most rational means to achieve the ends on which all reasonable people can agree" (2000a, 157). But Johnson doubts the power of reason. He thus rejects the inheritance of the Enlightenment by suggesting that we look elsewhere: "What we need is for God himself to speak, to give us a secure foundation on which we can build. If God has not spoken, then we have no alternative to despair. If God has spoken, then we need to build on that foundation rather than try to fit what God has done into some framework that comes from human philosophy" (2000a, 158). Johnson believes that God has indeed spoken, and the implications are potentially enormous.

The battles of the contemporary culture war in the United States and elsewhere may currently be waged over Darwinism, but as Johnson notes:

> Once we learn that nature does not really do its own creating, and we are not really products of mindless natural forces that care nothing about us, we will have to reexamine a great deal else. In particular, we will need to have a new discussion about the nature of *reason* . . . and what we might mean by the true, the good and the beautiful. Upon what foundation should we build our theories about all these things? After seeing that trying to build everything on a foundation of matter has led us into a blind alley, we will have to look for something better. Once the question is put that way, *Christians have an answer*. Scientists as such do not. (2000a, 159, my italics)

Put this way, a lot is indeed at stake. But as I noted earlier in this chapter, Christians do not have one answer; they have many divergent answers. Theirs is a diverse community. Many contradictory voices speak in the name of God. If reason itself is to be subordinate to God, we have no way to adjudicate these differences of opinion, since there will be no rational way to weigh and evaluate arguments and counterclaims that at least one of the parties to the debate will not object to as resting on a mistaken conception of God.

I agree that the way we see the world shapes our attitudes toward issues of value, but I suggest that seeing the world in methodologically naturalistic terms provides a much safer route to humane and decent values. Conservative Christianity may well flourish if we clip the wings of reason and inhibit rational inquiry into the nature of the world we live in, but it is not so clear that human beings will flourish under such a regime. Recent religiously motivated political debates about science education, prayer in schools, the posting of the Ten Commandments in public buildings, stem cell research, cloning, and the use of human embryonic tissue in biomedical research give us a mere foretaste of what is to come here.

It is very easy to support policies inhibiting research with potentially enormous benefits to human health and well-being if you believe that what really matters is a disembodied life after death—a claim that will be hard to challenge when the very standards of rational inquiry have been so crippled as to permit (and possibly mandate) widespread belief in the evidentially ungrounded supernatural.

In the end, it is the moral implications of the intelligent design movement and not its instrumental manipulation of science and science education that we all, collectively, need to be concerned about. For as Phillip Johnson has observed in his defense of the wedge strategy: "If reason is to be a reliable guide, it must be grounded in a foundation that is more fundamental than logic. . . . Instrumental reason is not enough. That is why the *fear of the Lord* is not the beginning of superstition but the beginning of wisdom" (2000a, 176, my italics). When reason rooted in evidence about the world is abandoned, we will all have much to fear, especially from those claiming to be the anointed worldly authorities of a supernatural being, the existence of which is rationally questionable. There is little doubt that the Lord's self-appointed worldly authorities are likely to flourish all too well in communities and societies where the citizenry won't be permitted to challenge their "wisdom" and will, like their ignorant medieval forebears prior to the Enlightenment, lack the intellectual tools and other means to do so.

Glossary

Adaptation Traditionally, adaptations were functional properties of organisms that were believed to reflect intelligent design. In this sense, organisms were viewed as being deliberately designed to fit into their appointed places in nature. Darwin showed that these same features of organisms could be explained as properties that enhanced reproductive success (and were observed in nature for this reason) and that were thus the fruits of the operation of the unintelligent, blind mechanism of natural selection.

Allele Several versions of a given gene may exist in a population. These different versions of a gene are known as *alleles*. Alleles typically differ from each other with respect to effects on the phenotypes of organisms possessing them. When these differences have implications for fitness, the alleles can become subject to the operation of natural selection.

Anthropic coincidences Our current theories about cosmology contain constants whose values are not predicted by theory and must be determined empirically. It is believed that if these constants had values different from the ones that have been empirically determined, life as we know it would not be possible. Some theorists see the specific values of the constants as coincidences requiring an explanation since they seem to reflect fine-tuning for life to have emerged; other theorists think the constants reflect pure chance.

Anthropic principles Interpretative principles, ranging from the sublime to the ridiculous, put forward to explain or otherwise make sense of the anthropic coincidences.

Argument from design An ancient argument for the existence of God(s) that proceeds from the observation of features of the world that seem inexplicable in terms of the operation of blind natural mechanisms to the conclusion that they result from deliberate intelligent design. In the context of Christianity, these arguments are taken to support the existence of God, viewed as a supernatural creator and/or designer. The cosmological versions of this argument typically focus on aspects of the physical world that seem to reflect deliberate design (e.g., the anthropic coincidences). Biological versions of this argument have focused on adaptations as features of organisms reflecting intelligent design. See also *Wedge strategy*.

Autocatalysis Autocatalysis occurs when the very presence of something promotes the formation of more of itself. Autocatalytic processes are studied in physics, chemistry, and biology.

Big bang According to modern cosmology, the universe emerged from a primal singularity about fourteen billion years ago. This event, leading to the expansion of the universe, is referred to as the big bang.

Catalyst Something that accelerates a reaction but remains unchanged in the process. In chemistry, substrates (initial substances) are often converted into products (final substances) with the help of catalysts.

Closed system See *Thermodynamics* and *Isolated system*.

Complex specified information (CSI) A type of information that some intelligent design theorists believe can be found in nature (for example, in irreducible complexity in biochemistry or in the anthropic coincidences in cosmology) and that is claimed to be the result of deliberate intelligent design. Scientists dispute the claim that the features of nature said to manifest CSI actually reflect the operation of a designing intelligence.

Cosmology The study of the nature of the universe, the way it changes over time, and the nature of its origins.

Creation science A branch of pseudoscience that attempts to support creationist claims mainly by attacking sciences such as physics, chemistry, biology, and geology. The attacks on natural science are pseudoscientific primarily because they rest on misinterpretations and misrepresentations (often willful and deliberate) of scientific claims and arguments. Intelligent design theory is an evolutionary descendant of creation science.

Creationism The religious theory that the universe, along with its biological contents, was created from nothing by a supernatural being. Young Earth creationists believe that God created the world about 6,000 years ago. Many believe humans and dinosaurs lived together. Old Earth creationists believe that the world, though created, is ancient.

Demiurge A hypothetical being spoken of in certain religious traditions whose role was to shape or form matter, much as an artisan or craftsman might shape or form artifacts from preexisting matter. The demiurge was not a creator of matter itself, merely a shaper.

Duplication A genetic process by means of which organisms get more genetic material. Duplications are very important in the context of evolution. A gene duplication results in an organism's genome containing an extra copy of a gene. The old copy of the gene can continue its function while the extra copy can be co-opted through exaptation to serve novel functions. Duplications are an important source of the redundant complexity that we see in organisms. An entire genome can be duplicated. This has been documented and has resulted in speciation in a single generation. These latter duplications of entire genomes are discussed under the heading of polyploidy. They are known to be important for speciation in plants, fish, amphibians, and insects.

Energy The total energy of a system is the sum of its potential energy and its kinetic energy. Potential energy is energy due to the location of a system, and kinetic energy is energy due to the motion of a system. For example, a ball at the top of a hill in the Earth's gravitational field has potential energy by virtue of its location. As the ball rolls down the hill, the potential energy is converted into kinetic energy. See also *First Law*.

Enlightenment Intellectual movement in the seventeenth and eighteenth centuries that was rooted in reason and science and a belief in the possibility of social progress and improvement. Enlightenment thinkers such as Hobbes, Locke, Rousseau, Voltaire, Tom Paine, and Thomas Jefferson challenged ideas in philosophy, politics, and religion that were based on tradition and authority. The thinkers of the Enlightenment challenged the idea of the divine right of kings and tended to champion democratic and other "popular" forms of government that were viewed as rational and progressive. The Enlightenment thinkers tended to be either deists (a view of religion based on reason) or atheists (thinkers critical of appeals to supernatural beings).

The first three presidents of the United States were deists. The emergence of democracy in the United States, as well as the great value placed on freedom and liberty, owes much to the Enlightenment critique of traditional forms of government, as does the idea that church and state must be separated. Intelligent design theory, along with extremist political and religious movements that seek to erode the wall of separation between church and state, can be viewed as a modern challenge to the ideals of the Enlightenment.

Entropy A quantity spoken of in thermodynamics. Intuitively, entropy is a measure of usable energy. Entropy increases as usable energy (energy available for work) decreases. A more detailed study of entropy shows that as entropy increases in a system, the system loses structural coherence; it tends to become more disordered. Under certain circumstances, when energy is allowed to flow through a system, the entropy of the system may decrease because of the phenomenon of self-organization.

Environment The environment can be broken into three components. The abiotic environment is the physical, nonbiological environment in which organisms are located. The biotic environment is the biological component of the environment, the predators, prey, pathogens, and parasites encountered by organisms. While organisms can change the physical environment, the biotic environment is actually evolving. Organisms thus coevolve together. A third component of the environment (perhaps a part of the biotic environment) is the cultural environment. Cultural effects, such as the invention of antibiotics, have implications for the evolutionary trajectories taken by biological populations, such as the bacteria that evolve antibiotic resistance.

Epistemology The study of the nature of knowledge and the rational justification of beliefs.

Equilibrium See *Thermodynamics*.

Evolution The observable phenomenon of change in the way populations of organisms are structured over time (usually, but not always, over many generations). Evolutionary mechanisms are the means by which populations of organisms change their structure over time. In modern terms, evolution occurs when, for any reason (not just natural selection), the statistical frequencies with which alleles are found in a population change. Microevolution is evolution below the species level. An example of microevolutionary change is afforded by the case of antibiotic resistance in the bacterial species *Staphylococcus aureus*. This is a bacterial adaptation brought about by natural selection. After many generations, natural selection has shaped populations of *S. aureus* to fit into environments (e.g., humans) periodically flooded with clinical doses of antibiotics. Microevolutionary changes, continued long enough, can bring about differences between distinct populations of a given species that are so great as to result in the formation of new species. The formation of new species and higher taxa are studied in the context of macroevolution. While speciation typically takes many generations (and can be directly studied in short-lived animals such as fruit flies), speciation through genome duplication can occur in a single generation.

Exaptation Process whereby an adaptation selected to serve a given function is co-opted or recruited by natural selection to serve a new function.

Fine-tuning Since life on Earth as we know it requires a universe with special properties (see also *Anthropic coincidences*), some theorists think that the values of the basic constants of physics have been finely tuned to make our existence possible—perhaps even to mandate it.

First Law A law of thermodynamics also known as the law of conservation of energy. Intuitively, it says that you cannot get something for nothing. More technically, it says that energy is neither created nor destroyed, though it can change its form (as when the chemical energy in a candle is turned into heat energy). More technically still, the law says that the total energy (sum of potential and kinetic energies) of a system isolated from external influences is constant.

Fitness A measure of the reproductive success of an organism. Characteristics of organisms that enhance their fitness will be found in higher frequencies in subsequent generations. Evolution occurs because different organisms leave behind different numbers of offspring. Adaptations are those characteristics of organisms that have enhanced fitness. See also *Natural selection*.

Fundamentalism While it is tempting to say that a fundamentalist is one who speaks from the fundament, the word is used in this book to apply to adherents of any religious tradition who claim that key religious statements in the tradition (e.g., that God made the world in six days) are literally, factually, and perhaps inerrantly true. Biblical literalism of this kind is a characteristic of Christian fundamentalists. Largely because of this, fundamentalists tend to be opaque to reason and immune to change of belief based on considerations of evidence (see also *creationism* and *creation science*). Fundamentalists need to be differentiated from traditionalists who adhere to traditional forms of religious expression (e.g., liturgical forms). Maintenance of traditional religious practices and rituals, along with support of traditional religious institutions, is thus not the same as fundamentalism, even if some traditionalists are also fundamentalists. Fundamentalists often use religion as a tool to advance social and political agendas (e.g., the "religious Right" in the United States, the Taliban in Afghanistan, and Islamic fundamentalists in terrorist groups such as Al Qaeda).

Gene A genetic unit situated at a particular locus or position on a chromosome. Structural genes make proteins (e.g., enzymes) that perform many roles in the life of an organism. Regulator genes control the expression of structural genes. Humans (and other diploid creatures) get two copies of a given gene, one from each parent. These copies of a gene are known as alleles. It is alleles (in sperm and egg) that pass through the bottleneck separating one generation from the next. Reproductive success can be measured in terms of an organism's contribution of alleles to the next generation. Alleles that enhance an organism's reproductive success will be found in higher frequencies in subsequent generations. Evolution occurs in a population when the frequencies with which alleles are found in a population undergo changes.

Genotype The genetic makeup of an organism. This can be thought of in terms of the combination of alleles it possesses. See *Phenotype*.

God In this book, the term refers to a hypothetical supernatural, supreme being, standing outside the physical world, who is believed by creationists to have both created the universe and to have intelligently designed the universe to contain living organisms, the adaptations of which are then viewed as the result of intelligent design (see *Argument from design*).

Intelligent design theory A theory based on the argument from design that evolved from creation science. Intelligent design theorists try to claim scientific status for their revivals of the argument from design. They reject many of the silly claims that made creation science laughable (e.g., that humans and dinosaurs once lived together, perhaps after the fashion of the Fred Flintstone cartoons). Intelligent design theorists tend to downplay the religious nature of their arguments as part of a wedge strategy to have religion taught in schools. The intelligent design movement is inextricably intertwined with extremist conservative politics and social agendas. See also *Natural theology* and *Wedge strategy*.

Irreducible complexity A type of complexity found in biochemistry that intelligent design theorists believe to be evidence of intelligent design. An irreducibly complex system has several components, all of which are needed for the system to function (and hence to have adaptive value). Since the absence of any component would lead the system to cease functioning, it is argued that if such a system was to evolve, it would have to evolve all at once, something that intelligent design theorists reject on the basis of sheer improbability. Irreducible complexity is viewed as a manifestation of complex specified information. Evolutionary theorists reject these claims, arguing instead that the key to the origin of irreducibly complex systems is redundant complexity.

Isolated system A system such that neither matter nor energy can cross the system boundary. Such a system is then said to be isolated from its surroundings. Because of terminological ambiguity, theorists sometimes refer to isolated systems as closed systems.

Literalism A view about the meaning of the Bible (or other texts) according to which the text has a literal meaning that is independent of the beliefs (and other cognitive features) of those who read and interpret it. In the case of the Bible, viewed as a consistent record of the true word of God, the literal meaning of the text was the one God had in mind. Literalists often disagree over what it was, exactly, that God had in mind. These problems are made all the more acute because the Bible seems to be logically inconsistent (Adam and Eve were created together in Genesis 1, whereas Adam was created prior to Eve in Genesis 2) and to contradict well-known mathematical truths (e.g., references in 1 Kings 7:23 to a circle 30 cubits in circumference and 10 in diameter, giving a false value of π—the ratio of circumference to diameter—as 3).

Metaphysics A branch of philosophy that deals with issues about the nature and characteristics of the objects in the universe and the constituents of reality. Materialism, for example, is the metaphysical theory according to which the only things that exist are physical objects and their properties. By contrast, supernaturalists believe that in addition to physical objects there also exist nonphysical, supernatural objects (e.g., souls, angels, demons, and Gods of various stripes). See also *Naturalism*.

Method of gradation A method in science for dealing with the relationships between things that appear to be very different from each other. Do the different objects or properties belong to separate categories or are they instead merely extreme points on a spectrum—one that is linked by intermediate cases (e.g., electrical conductors and nonconductors are linked through a range of semiconductors)? Darwin used the method to examine the traditional idea that distinct species were categorically different from each other (resulting, perhaps, from special acts of creation). Darwin's argument has the conclusion that distinct species—which may look very different from each other—descend from common ancestors with subsequent evolutionary modification.

Morphology The study of form or shape.

Multiverse A hypothetical structure that contains our universe as a part. Our universe might be one of many possible universes, where these are components of the multiverse.

Mutation A change in the genetic material of a cell (arising from point mutations or base-pair changes, insertions, deletions, duplications, etc.). Mutations in the genetic material in reproductive cells (sperm or egg) are known as *germ-line mutations*. These changes can be inherited and are an important source of heritable variation in populations of organisms. They thus provide some of the grist for the mill of natural selection.

Natural selection A mechanism, discovered by Darwin, by means of which the frequencies with which alleles are found in populations of organisms change over time. When a population of organisms displays variation with respect to characteristics of its component organisms, where this variation in characteristics leads to differential reproductive success or fitness (i.e., different organisms leaving behind different numbers of offspring), and where the characteristics contributing to the reproductive success of the parent(s) are probably inherited by offspring, then alleles coding for those characteristics can be expected to increase in frequency in the population. In this way, population structure changes over time, and populations can cope with changing environments. Natural selection is thus the natural mechanism that brings about adaptations without a need for intelligent design.

Natural theology The theory according to which discoveries about the nature of the physical world we live in (e.g., discoveries about the origins of the universe—see *fine-tuning*; or discoveries about biological systems—see *Irreducible complexity*) contain evidence of supernatural intelligent design by God. Intelligent design theory, far from being a form of natural science (as its proponents disingenuously suggest), is in fact nothing more than old natural theology in fancy new biochemical and cosmological bottles.

Naturalism, metaphysical A philosophical theory according to which nature is all that there is. This type of naturalism excludes from consideration claims about the existence of supernatural objects.

Naturalism, methodological A lesson learned by scientists, on the basis of long experience, according to which natural phenomena almost certainly admit of explanations in natural terms. The methodological naturalist does not categorically rule out (as a metaphysical naturalist would) the possibility that a phenomenon might need a nonnatural, supernatural explanation. The methodological naturalist (on the basis of long experience with psychics, astrologers, mystics, religious hoaxers, frauds, and tricksters) will tend to look for alternative natural explanations of anomalous phenomena and demand high-quality, unambiguous evidence before taking seriously claims about the need for supernatural explanations.

Ontogeny The developmental trajectory or path taken by an organism in the course of its life cycle. Developmental biologists have a special interest in ontogenetic phenomena.

Open-dissipative systems Physical systems that have exchanges of matter and energy with their environments. Organisms and hurricanes are examples of open-dissipative systems. When driven by energy flows and taken away from a state of thermodynamical equilibrium, such systems can exhibit the phenomenon of self-organization in which the entropy of the system actually decreases. See *Second Law* and *Thermodynamics*.

Phenotype The physical manifestation of an organism that results from complex interactions between genes and environments in the course of organismal development. The phenotype can be studied at many distinct levels of description, including macromolecules, intracellular structures, cells, tissues, organs, and on up to the entire, organized structure, the organism itself. See *Genotype*.

Phylogeny The historical pattern of evolutionary relationships between organisms. Evolutionary biologists are interested in uncovering these relationships by reconstructing historical events such as speciation events. The evidence used may include anatomical or physiological data, molecular and genetic data, behavioral data, developmental data, and last, but by no means least, fossil data.

Redundant complexity A type of complexity exhibited by biological systems. Redundantly complex systems consist of several parts but achieve their function in such a way that one or more parts can be removed without completely disrupting the system. As parts are removed and the system loses redundancy, it can be gradually transformed, in a series of evolutionary steps, into a system that manifests irreducible complexity. For this reason (among others), scientists are not convinced that irreducible complexity is a feature of biological systems that could not have evolved and must have resulted instead from intelligent design. Redundancy has many sources, but at the genetic level, duplications are a main source.

Science A collection of human activities aimed at explaining the origins and nature of things we see in the world around us and the changes that these things undergo. Science proceeds through the construction of theories, which should be supported by high-quality evidence, to explain things we have seen (directly, or indirectly with the aid of instruments), and help us make predictions about the future course of events. In historical sciences, such as cosmology, geology, linguistics, anthropology, and evolutionary biology, the aim of the activity is to construct theories deeply rooted in evidence that enable us to explain and understand the past.

Second Law A law of thermodynamics known as the *entropy law*. According to the Second Law, in an isolated system the energy available for work tends to a minimum and the entropy of the system tends to a maximum. Open-dissipative systems, taken away from equilibrium, may undergo self-organization, resulting in a decrease in entropy. In this case, the Second Law requires that the entropy reduction in the open-dissipative system be more than offset by an increase in the entropy of the environment with which it interacts, so that, on average, the total entropy of the combined system (open-dissipative system plus environment) increases.

Self-organization A dynamical phenomenon observed in many branches of science resulting in the formation of complex, organized systems with many interacting parts. The parts of such systems organize themselves without the aid of designing intelligences. Flows of energy activate interaction mechanisms, causing the components of such systems to form organized wholes that are typically greater than the sums of their parts.

Singularity A region of spacetime where the density of mass-energy, hence the curvature of spacetime, is infinite. Singularities are believed to lie at the hearts of black holes. According to the big bang theory of the origin of the universe, our universe (spacetime and mass-energy) emerged and continues to unfold from a singularity.

Supernatural A domain of objects spoken of in certain types of metaphysical theory, according to which there exist nonphysical beings standing outside of nature. The God of Christianity, along with such exotica as souls, demons, and angels, is typically conceived of as belonging to the supernatural domain, sometimes referred to as "the realm invisible." Scientists question whether such objects exist, not merely because we lack credible direct evidence of their existence, but because the indirect effects in nature that such objects are supposed to be responsible for (natural phenomena such as adaptations, recovery from illness, irreducible complexity, and, in the case of demons, mental illness) admit of simpler explanations in terms of purely natural causes—explanations for which there is abundant, high-quality evidence. Currently, there is no credible evidence of a direct or indirect nature to support the existence of supernatural objects or supernatural causes for natural effects.

Teleology The study of purposes, ends, goals, or functions. Intelligent design theory is a teleological theory in that it sees organisms and their parts (e.g., biochemical pathways) as being deliberately designed to serve various purposes and functions. In many branches of Christianity, it is believed that humans were created with a purpose in mind. Some Christians believe that engagement in certain activities or practices (e.g., homosexuality, abortion, tolerance of minorities, and even welfare programs for the poor) thwart the purposes for which we were designed. Such activities are said to be "unnatural" and are viewed as candidates for moral, social, and perhaps even legal sanction. Part of the impact of the theory of evolution is that it showed how features of organisms that appeared to be intelligently designed with a purpose in mind could result instead from the operation of the blind, natural mechanism of natural selection.

Theocracy A form of government in which political legitimacy is said to flow from a divine mandate. It may involve rule by religious authorities or rule by those who claim to be sanctioned by a higher, supernatural power, as in the "divine right of kings." The Enlightenment saw the end of theocracy in many Western nations (the Vatican is an exception). Iran is a good example of a modern theocracy, as was Afghanistan under the Taliban. Some intelligent design theorists seem to favor the modification of social and political institutions in the United States to be more theocratic in nature. The separation of church and state is a political and legal device to preserve democracy from degenerating into a theocracy. In a democracy, the legitimacy of government is derived from a mandate from the people (i.e., the voters) and not from the alleged demands and dictates of supernatural beings.

Theory A claim or collection of claims put forward to explain something we find puzzling. Theories may be speculative—for example, the claim that cattle mutilations are caused by aliens from outer space. Intelligent design theory is another example of a speculative theory. Such theories are not well supported by the evidence. But theories may be very well supported by lots of high-quality evidence. The central theories of modern science, such as the quantum theory, the theory of relativity, and the theory of evolution, are examples of such evidentially well-supported theories.

Thermodynamics The branch of science dealing with energy and its effects. In equilibrium thermodynamics, much of the interest is with isolated systems tending to a state of thermodynamical equilibrium in accord with the First Law and the Second Law. An isolated system can be approximated by a well-insulated box containing cold air and a hot lump of iron. As time goes by, the iron gets cooler and the air gets warmer until a state of thermal equilibrium is achieved, at which point the iron and the air are at the same temperature. Nonequilibrium thermodynamics considers what happens to open-dissipative systems when they are taken away from a condition of equilibrium with their surroundings. As matter and energy flow through such systems, self-organization can occur, bringing about a local reduction of entropy in accord with the Second Law. The Bénard cells discussed in chapter 3 are a good example of the phenomenon of self-organization as it occurs in the context of nonequilibrium thermodynamics.

Wedge strategy A political strategy favored by leading proponents of intelligent design theory cleverly designed to undermine science as we know it, as well as modern, secular forms of government. Wedge strategists seek to reject our common intellectual inheritance from the Enlightenment with a view to reinstating a theocratic form of government in which the wall of separation between church and state will no longer exist. A wedge has a thin end and a fat end. At the thin end of the wedge strategy are attempts at local, state, and national levels to get intelligent design theory taught alongside properly scientific theories about the world we live in. (God is rarely mentioned at this end of the wedge, and for this reason intelligent design theory has been referred to as "stealth creationism.") Toward the fat end of the wedge, a broader social and political agenda emerges, in which it becomes clear that intelligent design theory is a carefully crafted ideological vehicle intelligently designed to provide a theological justification for social causes close to the hearts of religious extremists.

Bibliography

Adair, R. K. 1987. *The Great Design: Particles, Fields and Creation*. Oxford: Oxford University Press.

Asimov, I. 1971. *The Universe: From Flat Earth to Quasar*. Baltimore: Penguin.

Atkins, P. W. 1994. *The Second Law: Energy, Chaos and Form*. New York: W. H. Freeman.

Babloyantz, A. 1986. *Molecules, Dynamics and Life*. New York: Wiley.

Barrow, J. D., and F. J. Tipler. 1986. *The Anthropic Cosmological Principle*. Oxford: Oxford University Press.

Behe, M. J. 1996. *Darwin's Black Box: The Biochemical Challenge to Evolution*. New York: Free Press.

———. 1999. Foreword to *Intelligent Design: The Bridge between Science and Theology*, by W. A. Dembski. Downers Grove, Ill.: Intervarsity Press.

———. 2000. Self-Organization and Irreducibly Complex Systems: A Reply to Shanks and Joplin. *Philosophy of Science* 67:155–162.

———. 2001a. Reply to Shanks and Joplin. *Reports of the National Center for Science Education* 21(4): 15.

———. 2001b. Darwin's Breakdown: Irreducible Complexity and Design at the Foundations of Life. In *Signs of Intelligence: Understanding Intelligent Design*, edited by W. A. Dembski and J. M. Kusiner. Grand Rapids, Mich.: Brazos Press.

Bennett, W. S. Jr., and T. A. Steitz. 1978. Glucose-Induced Conformational Changes in Yeast Hexokinase. *Proceedings of the National Academy of Sciences, USA* 75:4848–4852.

Berlocher, S. H. 1998. Origins: A Brief History of Research on Speciation. In *Endless Form: Species and Speciation*, edited by D. J. Howard and S. H. Berlocher. Oxford: Oxford University Press.

Bonabeau, E., M. Dorigo, and G. Theraulaz. 1999. *Swarm Intelligence: From Natural to Artificial Systems*. Oxford: Oxford University Press.

Boston Globe Investigative Staff. 2002. *Betrayal: The Crisis in the Catholic Church*. New York: Little, Brown.

Bradley, W. A. 2001. The "Just So" Universe: The Fine Tuning of Constants and Conditions in the Cosmos. In *Signs of Intelligence: Understanding Intelligent Design*, edited by W. A Dembski and J. M. Kusiner. Grand Rapids, Mich.: Brazos Press.

Bugge, T. H., K. W. Kombrinck, M. J. Flick, C. D. Daugherty, M. J. Danton, and J. L. Degan. 1996. Loss of Fibrinogen Rescues Mice for the Pleiotropic Effects of Plasminogen Deficiency. *Cell* 87:709–719.

Burnet, T. [1691] 1965. *The Sacred Theory of the Earth*. Carbondale, Ill.: SIU Press.

Butts, R. E. 1989. *William Whewell: Theory of Scientific Method*. Indianapolis: Hackett.

Cairns-Smith, A. G. 1986. *Seven Clues to the Origins of Life: A Scientific Detective Story*. Cambridge: Cambridge University Press.

Campbell, N. A. 1996. *Biology*: Menlo Park, Calif.: Benjamin Cummings.

Carroll, S. B., J. K. Grenier, and S. D. Weatherbee. 2001. *From DNA to Diversity: Molecular Genetics and the Evolution of Animal Design*. Malden, Mass.: Blackwell Science.

Clendening, L., ed. 1960. *Sourcebook of Medical History*. New York: Dover.

Coleman, W. 1977. *Biology in the Nineteenth Century: Problems of Form, Function and Transformation*. Cambridge: Cambridge University Press.

Corey, M. A. 2001. *The God Hypothesis: Discovering Design in Our "Just Right" Goldilocks Universe*. Oxford: Rowman & Littlefield.

Cottingham, J. 1986. *Descartes*. Oxford: Basil Blackwell.

———. 1988. *The Rationalists*. Oxford: Oxford University Press

Crombie, A. C. 1959. *Medieval and Early Modern Science*. Vol. 2. New York: Doubleday Anchor.

Cziko, G. 1995. *Without Miracles: Universal Selection Theory and the Second Darwinian Revolution*. Cambridge, Mass.: MIT Press.

Darwin, C. [1839] 1989. *Voyage of the Beagle*. Edited by J. Browne and M. Neve. New York: Penguin.

———. [1859] 1970. *The Origin of Species*. Edited by P. Appleman. New York: W. W. Norton.

———. [1872] 1965. *The Expression of the Emotions in Man and Animals*. Chicago: University of Chicago Press.

———. [1871] 1896. *The Descent of Man, and Selection in Relation to Sex*. Vol. 1. New York: D. Appleton.

Darwin, F. (ed.) 1888. *The Life and Letters of Charles Darwin*. Vol. 1. New York: Appleton.

———. 1888. *The Life and Letters of Charles Darwin*. Vol. 2. New York: Appleton.

Davidson, E. H. 2001. *Genomic Regulatory Systems: Development and Evolution*. San Diego: Academic Press.

Davies, P. R. 1992. In Search of "Ancient Israel." *Journal for the Study of the Old Testament*, supplement series, 148. Sheffield, England: Sheffield Academic Press.

———. 1998. *Scribes and Schools: The Canonization of the Hebrew Scriptures*. Louisville, KY: Westminster John Knox Press.

Dawkins, M. S. 1998. *Through Our Eyes Only: The Search for Animal Consciousness*. Oxford: Oxford University Press.

Dawkins, R. 1989. *The Selfish Gene*. Oxford: Oxford University Press.

Dembski, W. A. 1999. *Intelligent Design: The Bridge between Science and Theology*. Downers Grove, Ill: Intervarsity Press.

———. 2001a. Introduction: What Intelligent Design Is Not. In *Signs of Intelligence: Understanding Intelligent Design*, edited by W. A. Dembski and J. M. Kusiner. Grand Rapids, Mich.: Brazos Press.

———. 2001b. Signs of Intelligence: A Primer on the Discernment of Intelligent Design. In W. A. Dembski and J. M. Kusiner, eds. *Signs of Intelligence: Understanding Intelligent Design*. Grand Rapids, Mich.: Brazos Press.

———. 2002. *No Free Lunch: Why Specified Complexity Cannot Be Purchased without Intelligence*. New York: Rowman and Littlefield.

Dembski, W. A., and J. M. Kusiner, eds. 2001. *Signs of Intelligence: Understanding Intelligent Design*. Grand Rapids, Mich.: Brazos Press.

Depew, D. J., and B. H. Weber. 1995. *Darwinism Evolving: Systems Dynamics and the Genealogy of Natural Selection*. Cambridge, Mass.: MIT Press.

Dick, O. L., ed. 1978. *Aubrey's Brief Lives*. London: Penguin.

Dixit, A. K., and B. J. Nalebuff. 1991. *Thinking Strategically: The Competitive Edge in Business, Politics and Everyday Life*. New York: W. W. Norton.

Dixon, R. 1980. *Dynamic Astronomy*. Englewood Cliffs, N.J.: Prentice-Hall.

Dowehower, L. A., M. Harvey, B. L. Slagle, M. J. McArthur, C. A. Montgomery Jr., J. S. Butel, and A. Bradley. 1992. Mice Deficient for p53 Are Developmentally Normal but Susceptible to Spontaneous Tumors. *Nature* 356:215–221.

Dugatkin, L. A. 1997. *Cooperation among Animals: An Evolutionary Perspective*. Oxford: Oxford University Press.

Edis, T. 1999. Cloning Creationism in Turkey. *Reports of the National Center for Science Education* 19(1):30–35.

Elledge, R. M., and W.-H. Lee. 1995. Life and Death by p53. *BioEssays* 17:923–930.

Ellis, G. F. R., and R. M. Williams. 1988. *Flat and Curved Space-Times*. Oxford: Clarendon Press.

Elsberry, W. R. 1999. Review of the Design Inference. *Reports of the National Center for Science Education* 19(2):32–35.

Espinoza, N. R., and M. A. F. Noor. 2002. Population Genetics of a Polyploid: Is There Hybridization between Lineages of Hyla versicolor? *Journal of Heredity* 93:81–85.

Ewald, P. W. 1994. *The Evolution of Infectious Disease*. Oxford: Oxford University Press.

Fairweather, A. M. 1954. *Nature and Grace: Selections from the Summa Theologica of Thomas Aquinas*. Vol. 11. Philadelphia: Westminster Press.

Finkelstein, I., and N. A. Silberman. 2001. *The Bible Unearthed: Archaeology's New Vision and the Origin of Its Sacred Texts*. New York: Free Press.

Futuyma, D. 1998. *Evolutionary Biology*. Sunderland, Mass.: Sinauer.

Gale, G. 1981. The Anthropic Principle. *Scientific American* 245:154–171.

———. 1990. Cosmological Fecundity: Theories of Multiple Universes. In *Physical Cosmology and Philosophy*, edited by J. Leslie. Dordrecht: Reidel.

———. 1997. Anthropic Principle: Physics or Metaphysics? In *Final Causality in Nature and Human Affairs*, edited by R. F. Hassing. Washington, D.C.: Catholic University of America Press.

Gale, G., and N. Shanks. 1997. Methodology and the Birth of Modern Cosmology. *Studies in the History and Philosophy of Modern Physics* 27:279–296.

Geach, P. T. 1979. *The Virtues*. Cambridge: Cambridge University Press.

Gerhart, J., and M. Kirschner. 1997. *Cells, Embryos and Evolution: Toward a Cellular and Developmental Understanding of Phenotypic Variation and Evolutionary Adaptability*. Oxford: Blackwell.

Goodwin, B. 1996. *How the Leopard Changed Its Spots: The Evolution of Complexity*. New York: Harper Torchbooks.

Gordon, B. L. 2001. Is Intelligent Design Science? The Scientific Status and Future of Design-Theoretic Explanations. In *Signs of Intelligence: Understanding Intelligent Design*, edited by W. A. Dembski and J. M. Kusiner. Grand Rapids, Mich.: Brazos Press.

Gould, S. J. 1981. *The Mismeasure of Man*. New York: W. W. Norton.

———. 1987. *Time's Arrow, Time's Cycle: Myth and Metaphor in the Discovery of Geological Time*. Cambridge, Mass.: Harvard University Press.

———. 1993. *Eight Little Piggies: Reflections in Natural History*. New York: W. W. Norton.

———. 1995. *Dinosaur in a Haystack: Reflections in Natural History*. New York: Crown.

Greaves, M. 2000. *Cancer: The Evolutionary Legacy*. Oxford: Oxford University Press.

Guth, A., and P. Steinhardt. 1989. The Inflationary Universe. In *The New Physics*, edited by P. C. W. Davies. Cambridge: Cambridge University Press.

Harris, W. S., H. Gowda, J. W. Kolb, C. P. Strychacz, J. L. Vacek, P. G. Jones, A. Farker, J. H. O'Keefe, and B. D. McCallister. 1999. A Randomized, Controlled Trial of Effects of Remote, Intercessionary Prayer and Outcomes in Patients Admitted to the Cardiac Care Unit. *Archives of Internal Medicine* 159:2273–2278.

Hawking, S. 1988. *A Brief History of Time: From the Big Bang to Black Holes*. New York: Bantam.

———. 1989. The Edge of Spacetime. In *The New Physics*, edited by P. C. W. Davies. Cambridge: Cambridge University Press.

Hochachka, P. W., and G. N. Somero. 2002. *Biochemical Adaptation: Mechanism and Process in Physiological Evolution*. Oxford: Oxford University Press.

Hogan, C. J. 1998. *The Little Book of the Big Bang*. New York: Springer-Verlag.

———. 2002. Observing the Beginning of Time. *American Scientist* 90:420–427.

Hope, J. 1996. A Better Mousetrap. *American Heritage* (October):90–97.

Hornell, J. 1940. Old English Dead-Fall Traps. *Antiquity* 14:395–403.

Hume, D. [1777] 1975. *Enquiries concerning Human Understanding and concerning the Principles of Morals*. Oxford: Clarendon Press.

Johnson, P. E. 1993. *Reason in the Balance: The Case against Naturalism in Science, Law and Education*. Downers Grove, Ill: Intervarsity Press.

———. 1997. *Defeating Darwinism by Opening Minds*. Downers Grove, Ill.: InterVarsity Press.

———. 2000a. *The Wedge of Truth: Splitting the Foundations of Naturalism*. Downers Grove, Ill: Intervarsity Press.

———. 2000b. Author's Response. *Metascience* 9(1):102–107.

———. 2001. The Intelligent Design Movement. In *Signs of Intelligence: Understanding Intelligent Design*, edited by W. A. Dembski and J. M. Kusiner. Grand Rapids, Mich.: Brazos Press.

Karsai, I., and Z. Penzes. 1998. Nest Shapes in Paper Wasps: Can the Variability of Forms Be Deduced from the Same Construction Algorithm. *Proceedings of the Royal Society of London*. B. 256:1261–1268.

———. 2000. Optimality of Cell Arrangement and Rules of Thumb of Cell Initiation in *Polistes dominulus:* A Modeling Approach. *Behavioral Ecology* 11:387–395.

Kauffman, S. A. 1993. *The Origins of Order: Self-Organization and Selection in Evolution*. Oxford: Oxford University Press.

———. 1995. *At Home in the Universe: The Search for the Laws of Self-Organization and Complexity*. Oxford: Oxford University Press.

Keller, E. F., and E. A. Lloyd, eds. 1992. *Keywords in Evolutionary Biology*. Cambridge, Mass.: Harvard University Press.

Knapp, R. G., and M. C. Miller. 1992. *Clinical Epidemiology and Biostatistics*. Malvern, Pa.: Harwal.

Koenig, H. G., K. I. Pargament, and J. Nielsen. 1998. Religious Coping and Health Status in Medically Ill Hospitalized Older Adults. *Journal of Nervous and Mental Disease* 186(9):513–521.

LaFollette, H. L., and N. Shanks. 1996. *Brute Science: The Dilemmas of Animal Experimentation*. London: Routledge.

Lahav, N. 1999. *Biogenesis: Theories of Life's Origin*. Oxford: Oxford University Press.

Land, M. F., and D.-E. Nilsson. 2002. *Animal Eyes*. Oxford: Oxford University Press.

Lehninger, A., D. Nelson, and M. Cox. 1993. *Principles of Biochemistry*. New York: Worth.

Lewis, R. 1995. Apoptosis Activity: Cell Death Establishes Itself as a Lively Research Field. *Scientist* 9:15.

Lewontin, R. C. 1992. Genotype and Phenotype. In *Keywords in Evolutionary Biology*, edited by E. F. Keller and E. A. Lloyd. Cambridge, Mass.: Harvard University Press.

———. 1995. Primate Models of Human Traits. *Perspectives on Medical Research* 5:5–19.

———. 2000. *The Triple Helix: Gene, Organism and Environment*. Cambridge, Mass.: Harvard University Press.

Li, W.-H. 1997. *Molecular Evolution*. Sunderland, Mass.: Sinauer.

Lyell, C. 1861. *Principles of Geology or the Modern Change of the Earth and Its Inhabitants Considered as Illustrative of Geology*. Vol. 1. London: John Murray.

———. 1861. *Principles of Geology or the Modern Change of the Earth and Its Inhabitants Considered as Illustrative of Geology*. Vol. 2. London: John Murray.

Macphail, E. M. 1998. *The Evolution of Consciousness*. Oxford: Oxford University Press.

Malthus, T. R. [1798] 1966. *First Essay on Population*. New York: St. Martin's Press.

Martini, G., and M. V. Ursini. 1996. A New Lease of Life for an Old Enzyme. *BioEssays* 18:631–637.

Mason, S. F. 1962. *A History of the Sciences*. New York: Collier.

Maudlin, T. 2002. *Quantum Non-Locality and Relativity*. Oxford: Blackwell.

Maynard Smith, J. 2000. *Evolutionary Genetics*. Oxford: Oxford University Press.

Maynard Smith, J., and E. Szathmáry. 1999. *The Origins of Life: From the Birth of Life to the Origins of Language*. Oxford: Oxford University Press.

Mayr, E. 1988. *Toward a New Philosophy of Biology: Observations of an Evolutionist*. Cambridge, Mass.: Harvard University Press.

———. 1991. *One Long Argument: Charles Darwin and the Genesis of Modern Evolutionary Thought*. Cambridge, Mass.: Harvard University Press.

———. 1997. *This Is Biology: The Science of the Living World*. Cambridge, Mass.: Harvard University Press.

Meiklejohn, J. M. D. 1969. *Immanuel Kant: Critique of Pure Reason*. New York: Dutton.

Mill, J. S. 1972. *Utilitarianism, On Liberty, and Considerations of Representative Government*. London: J. M. Dent.

Miller, K. R. 1999. *Finding Darwin's God: A Scientist's Search for Common Ground between God and Evolution*. New York: Harper Collins.

Morris, H. 1974. *The Troubled Waters of Evolution*. San Diego: Creation Life.

Murray, A., and T. Hunt. 1993. *The Cell Cycle*. Oxford: Oxford University Press.

Nesse, R. M., and G. C. Williams. 1995. *Why We Get Sick: The New Science of Darwinian Medicine*. New York: Vintage.

Paley, W. [1801] 1850. *Natural Theology, or Evidence of the Existence and Attributes of the Deity, Collected from the Appearances of Nature*. New York: American Tract Society.

Pargament, K. I., H. G. Koenig, N. Tarakeshwar, and J. Hahn. 2001. Religious Struggle as a Predictor of Mortality among Medically Ill Elderly Patients. *Archives of Internal Medicine* 161(13): 1881–1884.

Parham, P. 1994. The Rise and Fall of Great Class I Genes. *Seminars in Immunology*, 6.

———. 2000. *The Immune System*. New York: Garland.

Park, M. A. 1996. *Biological Anthropology*. Mountain View, Calif.: Mayfield.

Parkinson, G. H. R., ed. 1977. *Leibniz: Philosophical Writings*. London: J. M. Dent and Sons.

Pearcey, N. 2001. Design and the Discriminating Public: Gaining a Hearing from Ordinary People. In *Signs of Intelligence: Understanding Intelligent Design*, edited by W. A. Dembski and J. M. Kusiner. Grand Rapids, Mich.: Brazos Press.

Perakh, M. 2003. *Unintelligent Design*. New York: Prometheus.

Petroski, H. 1992. *To Engineer Is Human: The Role of Failure in Successful Design*. New York: Vintage.

———. 1994. *The Evolution of Useful Things*. New York: Vintage.

Pigliucci, M. 2003. Species as Family Resemblance Concepts: The (Dis-) Solution of the Species Problem. Forthcoming in *BioEssays*.

Pike, N. 1970. *Hume: Dialogues concerning Natural Religion*. New York: Bobbs-Merrill.

Powers, J. 1982. *Philosophy and the New Physics*. London: Methuen.

Price, P. W. 1996. *Biological Evolution*. New York: Saunders.

Rachels, J. 1991. *Created from Animals: The Moral Implications of Darwinism*. Oxford: Oxford University Press.

Rackham, H. 1979. *Cicero: De Natura Deorum Academica*. Cambridge, Mass.: Harvard University Press.

Rees, M. 1997. *Before the Beginning: Our Universe and Others*. Cambridge, Mass.: Perseus.

———. 1999. *Just Six Numbers: The Deep Forces That Shape the Universe*. New York: Basic Books.

———. 2001. *Our Cosmic Habitat*. Princeton, N.J.: Princeton University Press.

Regan, T., and P. Singer, eds. 1989. *Animal Rights and Human Obligations*. Englewood Cliffs, N.J.: Prentice Hall.

Ridley, M. 1998. *The Origin of Virtue: Human Instinct and the Evolution of Cooperation*. New York: Viking.

Rosenzweig, M. L. 1995. *Species Diversity in Space and Time*. Cambridge: Cambridge University Press.

Ross, H. 2001. *The Creator of the Cosmos: How the Greatest Scientific Discoveries of the Century Reveal God*. Colorado Springs, Colo.: NavPress.

Rossum, G. D.-v. 1996. *History of the Hour: Clocks and Modern Temporal Orders*. Chicago: University of Chicago Press.

Sayin, Ü., and A. Kence. 1999. Islamic Scientific Creationism: A New Challenge in Turkey. *Reports of the National Center for Science Education* 19(6):18–29.

Seeley, T. D. 2002. When Is Self-Organization Used in Biological Systems? *Biological Bulletin* 202:314–318.

Shallit, J. 2002. Review of *No Free Lunch* by William Dembski. *BioSystems* 66:93–99.

Shanks, N. 1991. Probabilistic Physics and the Metaphysics of Time. *South African Journal of Philosophy* 10(2):37–43.

———. 1993. Axiomatic Quantum Mechanics and Radioactive Decay. *Erkenntnis* 39:243–255.

———. 1994. Time, Physics and Freedom. *Metaphilosophy* 25(1):45–59.

———. 1995. Minds, Brains and Quantum Mechanics. *Southern Journal of Philosophy* 33(2):243–260.

———. 2000. Creationism, Evolution and Baloney. *Metascience* 9(1):86–93.

———. 2001a. Fighting for Our Sanity in Tennessee: Life on the Front Lines. *Free Inquiry* 21(4):38–40.

———. 2001b. Modeling Biological Systems: The Belousov-Zhabotinski Reaction. *Foundations of Chemistry* 3:33–53.

———. 2002. *Animals and Science: A Guide to the Debates*. Santa Barbara, Calif.: ABC-Clio.

Shanks, N., and K. H. Joplin. 1999. Redundant Complexity: A Critical Analysis of Intelligent Design in Biochemistry. *Philosophy of Science* 66:268–282.

———. 2000. Of Mousetraps and Men: Behe on Biochemistry. *Reports of the National Center for Science Education* 20:25–30.

———. 2001a. Behe, Biochemistry and the Invisible Hand. *Philo* 4:54–67.

———. 2001b. Reply to Behe. *Reports of the National Center for Science Education* 21(4):16.

Shapiro, A. 1999. Fundamentalist Bedfellows: Political Creationism in Turkey. *Reports of the National Center for Science Education* 19(6):15–20.

Shettleworth, S. J. 1998. *Cognition, Evolution and Behavior*. Oxford: Oxford University Press.

Sinclair, T. A. 1976. *Aristotle: The Politics*. Middlesex, England: Penguin.

Sloan, R. P., and E. Bagiella. 2000. Data without a Prayer. *Archives of Internal Medicine* 160:1870.

Sloan, R. P., E. Bagiella, and T. Powell. 1999. Religion, Spirituality and Medicine. *Lancet* 353:664–667.

Sober, E. 2000. *Philosophy of Biology*. Boulder, Colo.: Westview Press.

Solé, R., and B. Goodwin. 2000. *Signs of Life: How Complexity Pervades Biology*. New York: Basic Books.

Stenger, V. J. 1998. Anthropic Design and the Laws of Physics. *Reports of the National Center for Science Education* 18(3):8–12.

———. 2000. Natural Explanations for the Anthropic Coincidences. *Philo* 3(2):50–67.

Strahler, A. N. 1987. *Science and Earth History: The Evolution/Creation Controversy*. New York: Prometheus.

Stumpf, S. E. 1982. *From Socrates to Sartre: A History of Philosophy*. New York: McGraw-Hill.

Templeton, A. R. 1989. The Meaning of Species and Speciation: A Genetic Perspective. In *Speciation and Its Consequences*, edited by D. Otte and J. A. Endler. Sunderland, Mass.: Sinauei.

Templeton, J. 2000. *Possibilities: The Humble Approach in Theology and Science*. Philadelphia: Templeton Foundation Press.

Thayer, H. S., ed. 1953. *Newton's Philosophy of Nature: Selections from His Writings*. New York: Hafner.

Thompson, T. L. 1999. *The Mythic Past: Biblical Archaeology and the Myth of Ancient Israel*. New York: Basic Books.

Tipler, F. J. 1995. *The Physics of Immortality: Modern Cosmology, God and the Resurrection of the Dead*. New York: Anchor.

Travis, J. 1992. Scoring a Technical Knockout in Mice. *Science* 256:1392–1394.

Tyson, J. T. 1994. What Everyone Should Know about the Belousov-Zhabotinski Reaction. In *Frontiers in Mathematical Biology*, edited by S. A. Levin. New York: Springer-Verlag.

Voet, D., and J. G. Voet. 1995. *Biochemistry*. New York: Wiley.

Wald, R. M. 1984. *General Relativity*. Chicago: University of Chicago Press.

Warnock, G. J. 1971. *The Object of Morality*. London: Methuen.

Weinberg, S. 1988. *The First Three Minutes: A Modern View of the Origin of the Universe*. New York: Basic Books.

West, J. G. 2001. The Regeneration of Science and Culture: The Cultural Implications of Scientific Materialism versus Intelligent Design. In *Signs of Intelligence: Understanding Intelligent Design*, edited by W. A. Dembski and J. M. Kusiner. Grand Rapids, Mich.: Brazos Press.

Wilkins, A. S. 2002. *The Evolution of Developmental Pathways*. Sunderland, Mass.: Sinauer.

Winny, J. 1965. *The General Prologue to the Canterbury Tales*. Cambridge: Cambridge University Press.

Wray, G. A. 2001. Development: Resolving the *Hox* Paradox. *Science* 292:2256–2257.

Yahya, H. 2001. *Evolution Deceit: The Scientific Collapse of Darwinism and Its Ideological Background*. Istanbul: Kültür.

Index